181 Totally Positive Matrices

Totally Positive Matrices

ALLAN PINKUS

Technion, Israel

CAMBRIDGE
UNIVERSITY PRESS

CAMBRIDGE UNIVERSITY PRESS
Cambridge, New York, Melbourne, Madrid, Cape Town,
Singapore, São Paulo, Delhi, Tokyo, Mexico City

Cambridge University Press
The Edinburgh Building, Cambridge CB2 8RU, UK

Published in the United States of America by
Cambridge University Press, New York

www.cambridge.org
Information on this title: www.cambridge.org/9780521194082

First published 2010

A catalogue record for this publication is available from the British Library

ISBN 978-0-521-19408-2 Hardback

This monograph is dedicated to the memory of
I. J. Schoenberg, M. G. Krein, F. R. Gantmacher and S. Karlin,
the four pioneers of the theory of total positivity.

We work in the dark – we do what we can – we give what we have.
Our doubt is our passion, and our passion is our task.

Contents

Foreword

In this monograph was present the central properties of finite totally positive matrices. As such, the monograph has only six main chapters. We consider the basic properties of such matrices, determinantal criteria for when a matrix is totally positive, their variation diminishing properties, various examples of totally positive matrices, their eigenvalue/eigenvector properties, and factorizations of such matrices. Numerous topics are excluded from this exposition. Total positivity is a theory of considerable consequence, and the most glaring omissions of this monograph are undoubtedly its various applications to diverse areas. Aside from the many applications mentioned in Gantmacher, Krein [1950] and Karlin [1968], applications can be found in approximation theory (see Schumaker [1981], Pinkus [1985c]), combinatorics (see Brenti [1989], [1995], [1996]), graph theory (see Fomin, Zelevinsky [2000], Berenstein, Fomin, Zelevinsky [1996]), Lie group theory (see Lusztig [1994]), majorization (see Marshall, Olkin [1979]), noncommutative harmonic analysis (see Gross, Richards [1995]), shape preservation (see Goodman [1995]), computing using totally positive matrices (see de Boor, Pinkus [1977], Koev [2005], Demmel, Koev [2005], Koev [2007]), refinement equations and subdivision (see Cavaretta, Dahmen, Micchelli [1991], Micchelli, Pinkus [1991]), and infinite totally positive banded matrices (see Cavaretta, Dahmen, Micchelli, Smith [1981], de Boor [1982], Smith [1983], Dahmen, Micchelli, [1986]). See also the many references in these papers and also the many references to these papers. There has been no attempt to make this monograph all-encompassing, and we apologize to all who feel that their contributions to the theory have been slighted as a consequence.

The theory of totally positive matrices is an odd bird in the matrix theory aviary. Much of the motivation for its study has come from problems in analysis, and the main initiators and contributors to the theory were analysts. I. J. Schoenberg was interested in the problem of estimating the number of real zeros of a polynomial, and this led him to his work on variation diminishing transformations (in the early 1930s) and Pólya frequency sequences, functions, and kernels (late 1940s and early 1950s). These, together with his work on splines (1960s and 1970s), are central topics in the theory of total positivity. M. G. Krein was led to the theory of total positivity via ordinary differential equations whose Green's functions are totally positive (mid 1930s). S. Karlin came to the theory of total positivity (in the 1950s and 1960s) by way of statistics, reliability theory, and mathematical economics. The two major texts on the subject *Oscillation Matrices and Kernels and Small Vibrations of Mechanical Systems*, by F. R. Gantmacher and M. G. Krein (see Gantmacher, Krein [1950]), and *Total Positivity. Volume 1*, by S. Karlin (see Karlin [1968]), are a blend of analysis and matrix theory (and in the latter case the emphasis is most certainly on analysis). (Their companion volumes *The Markov Moment Problem and Extremal Problems*, by M. G. Krein and A. A. Nudel'man (see Krein, Nudel'man [1977]) and *Tchebycheff Systems: with Applications in Analysis and Statistics*, by S. Karlin and W. J. Studden (see Karlin, Studden [1966]), are totally devoted to topics of analysis.) Thankfully we have the short monograph of T. Ando that eventually appeared as Ando [1987] (it was written a few years earlier) and was devoted to totally positive matrices. The present monograph is an attempt to update and expand upon Ando's monograph. A considerable amount of research has been devoted to this area in the past twenty years, and such an update is certainly warranted.

It was Schoenberg, in Schoenberg [1930], who coined the term *total positiv* (in German). Krein and Gantmacher (see Gantmacher, Krein [1935]), unaware of Schoenberg's earlier paper, used the term *complètement non négative* and *complètement positive* (French) for totally positive and strictly totally positive, respectively. As such, many authors use the term *totally nonnegative* and *totally positive* for totally positive and strictly totally positive, respectively, which, aside from the lack of consistency and order, all too often leads to confusion. We follow the Schoenberg/Karlin/Ando terminology.

It is a pleasure to acknowledge the help of Carl de Boor and David Tranah. All errors, omissions and other transgressions are the author's responsibility.

I would like to close this short foreword with a personal note. My first mathematical paper (jointly written with my doctoral supervisor Sam Karlin) was in the area of total positivity. It is said that as one gets old(er) one often returns to one's first love. I plead guilty on both counts.

Haifa, 2008.

1

Basic properties of totally positive and strictly totally positive matrices

In this chapter, we introduce some of the notation, basic definitions, and various classic facts and formulæ. Many of the results of this chapter will be used in subsequent chapters. Matrix notation, more especially the notation used for submatrices and minors, is both clumsy and problematic. It definitely takes getting used to. But rest assured that one does eventually get used to it. The medium here is not the message.

In Section 1.2 we consider some simple and less simple operations that preserve total positivity and strict total positivity. Section 1.3 is about nonsingularity and rank. Here we immediately see results that are less than obvious. Finally, in Section 1.4, we present a few basic determinantal inequalities that are valid for totally positive and strictly totally positive matrices.

1.1 Preliminaries

For a positive integer n, and for each $p \in \{1, \ldots, n\}$, we define the simplex

$$I_p^n := \{\mathbf{i} = (i_1, \ldots, i_p) \, : \, 1 \le i_1 < \cdots < i_p \le n\}$$

in $\mathbb{Z}_+^p$. That is, I_p^n denotes the set of strictly increasing sequences of p integers in $\{1, \ldots, n\}$.

We use the following notation to define submatrices and minors of a matrix. If $A = (a_{ij})_{i=1}^{n}{}_{j=1}^{m}$ is an $n \times m$ matrix, then for each $\mathbf{i} \in I_p^n$ and $\mathbf{j} \in I_q^m$ we let

$$A[\mathbf{i}, \mathbf{j}] = A \begin{bmatrix} \mathbf{i} \\ \mathbf{j} \end{bmatrix} = A \begin{bmatrix} i_1, \ldots, i_p \\ j_1, \ldots, j_q \end{bmatrix} := (a_{i_k j_\ell})_{k=1}^{p}{}_{\ell=1}^{q}$$

denote the $p \times q$ submatrix of A determined by the rows indexed $i_1, \ldots, i_p$

and columns indexed $j_1, \ldots, j_q$. When $p = q$ then

$$A(\mathbf{i}, \mathbf{j}) = A\begin{pmatrix} \mathbf{i} \\ \mathbf{j} \end{pmatrix} = A\begin{pmatrix} i_1, \ldots, i_p \\ j_1, \ldots, j_p \end{pmatrix} := \det\,(a_{i_k j_\ell})_{k,\ell=1}^p$$

denotes the associated minor, i.e., the determinant of the submatrix.

This monograph is about totally positive and strictly totally positive matrices. They are defined as follows:

Definition 1.1 An $n \times m$ matrix A is said to be *totally positive* (TP) if all its minors are nonnegative, i.e.,

$$A(\mathbf{i}, \mathbf{j}) = A\begin{pmatrix} i_1, \ldots, i_p \\ j_1, \ldots, j_p \end{pmatrix} \geq 0 \tag{1.1}$$

for all $\mathbf{i} \in I_p^n$, $\mathbf{j} \in I_p^m$, and all $p = 1, \ldots, \min\{n, m\}$. It is said to be *strictly totally positive* (STP) if strict inequality always holds in (1.1).

We use and reuse various classic facts and formulæ. We list some of them here for easy reference.

Cauchy–Binet and pth compound matrices The pth *compound matrix* of the $n \times m$ matrix A is denoted by $A_{[p]}$ and is defined as the $\binom{n}{p} \times \binom{m}{p}$ matrix with entries

$$(A(\mathbf{i}, \mathbf{j}))_{\mathbf{i} \in I_p^n, \mathbf{j} \in I_p^m}$$

where the $\mathbf{i} \in I_p^n$ and $\mathbf{j} \in I_p^m$ are arranged in lexicographic order, i.e., for distinct $\mathbf{i}, \mathbf{k} \in I_p^n$ we set $\mathbf{i} > \mathbf{k}$ if the first nonzero term in the sequence $i_1 - k_1, \ldots, i_p - k_p$ is positive.

Assume $B = CD$, where B is an $n \times m$ matrix, C is an $n \times r$ matrix, and D is an $r \times m$ matrix. The *Cauchy–Binet formula* may be written as follows. For each $\mathbf{i} \in I_p^n$, $\mathbf{j} \in I_p^m$,

$$B(\mathbf{i}, \mathbf{j}) = \sum_{\mathbf{k} \in I_p^r} C(\mathbf{i}, \mathbf{k}) D(\mathbf{k}, \mathbf{j}),$$

i.e.,

$$B\begin{pmatrix} i_1, \ldots, i_p \\ j_1, \ldots, j_p \end{pmatrix} = \sum_{1 \leq k_1 < \cdots < k_p \leq r} C\begin{pmatrix} i_1, \ldots, i_p \\ k_1, \ldots, k_p \end{pmatrix} D\begin{pmatrix} k_1, \ldots, k_p \\ j_1, \ldots, j_p \end{pmatrix},$$

or, alternatively,

$$B_{[p]} = C_{[p]} D_{[p]}.$$

This is, of course, a generalization of the formula for matrix multiplication. For $p = 1$ the above reduces to

$$b_{ij} = \sum_{k=1}^{r} c_{ik} d_{kj}.$$

This Cauchy–Binet formula is valid when $p \le \min\{n, m, r\}$. For $p > r$ ($p \le \min\{n, m\}$) the set I_p^r is empty, and in the above sum we set $B(\mathbf{i}, \mathbf{j}) = 0$. This is the "correct convention" as $\operatorname{rank} B \le r$.

For p vectors $\mathbf{u}^1 = (u_{1,1}, \ldots, u_{n,1}), \ldots, \mathbf{u}^p = (u_{1,p}, \ldots, u_{n,p}) \in \mathbb{C}^n$, and each $\mathbf{i} \in I_p^n$ we set

$$\left(\mathbf{u}^1 \wedge \cdots \wedge \mathbf{u}^p\right)(\mathbf{i}) = \det\left(u_{i_\ell, j}\right)_{\ell, j=1}^{p}.$$

We consider $\mathbf{u}^1 \wedge \cdots \wedge \mathbf{u}^p$ as a vector in $\mathbb{C}^{\binom{n}{p}}$. It is termed the *Grassman product* or *wedge product* or *exterior product* of $\mathbf{u}^1, \ldots, \mathbf{u}^p$. Obviously $\mathbf{u}^1 \wedge \cdots \wedge \mathbf{u}^p = \mathbf{0}$ if and only if the $\mathbf{u}^1, \ldots, \mathbf{u}^p$ are linearly dependent. From the Cauchy–Binet formula it easily follows that

$$A_{[p]}(\mathbf{u}^1 \wedge \cdots \wedge \mathbf{u}^p) = A\mathbf{u}^1 \wedge \cdots \wedge A\mathbf{u}^p.$$

Sylvester's Determinant Identity Let A be an $n \times m$ matrix, and let

$$(\alpha_1, \ldots, \alpha_p) \in I_p^n \quad \text{and} \quad (\beta_1, \ldots, \beta_p) \in I_p^m.$$

For each $i \in \{1, \ldots, n\} \backslash \{\alpha_1, \ldots, \alpha_p\}$ and $j \in \{1, \ldots, m\} \backslash \{\beta_1, \ldots, \beta_p\}$ we set

$$b_{ij} = A\begin{pmatrix} k_1, \ldots, k_{p+1} \\ \ell_1, \ldots, \ell_{p+1} \end{pmatrix}$$

where $\{k_1, \ldots, k_{p+1}\}$ is the set of integers $\{\alpha_1, \ldots, \alpha_p, i\}$ arranged in natural (increasing) order, and $\{\ell_1, \ldots, \ell_{p+1}\}$ is the set of integers $\{\beta_1, \ldots, \beta_p, j\}$ arranged in natural order. We generally abuse notation by writing

$$b_{ij} = A\begin{pmatrix} \alpha_1, \ldots, \alpha_p, i \\ \beta_1, \ldots, \beta_p, j \end{pmatrix}.$$

But it is **always** to be understood that we have arranged these row and column indices in natural order.

Sylvester's Determinant Identity states that the minors of the $(n-p) \times (m-p)$ matrix $B = (b_{ij})$ satisfy

$$B\begin{pmatrix} i_1, \ldots, i_r \\ j_1, \ldots, j_r \end{pmatrix} = \left[A\begin{pmatrix} \alpha_1, \ldots, \alpha_p \\ \beta_1, \ldots, \beta_p \end{pmatrix}\right]^{r-1} A\begin{pmatrix} \alpha_1, \ldots, \alpha_p, i_1, \ldots, i_r \\ \beta_1, \ldots, \beta_p, j_1, \ldots, j_r \end{pmatrix}.$$

The submatrix

$$A\begin{bmatrix}\alpha_1,\ldots,\alpha_p\\ \beta_1,\ldots,\beta_p\end{bmatrix}$$

is called the *pivot block.*

Inverses If $A = (a_{ij})$ is an $n \times n$ nonsingular matrix, then the elements of its inverse $A^{-1} = (c_{ij})$ satisfy

$$c_{ij} = \frac{(-1)^{i+j} A\begin{pmatrix}1,\ldots,\widehat{j},\ldots,n\\ 1,\ldots,\widehat{i},\ldots,n\end{pmatrix}}{A\begin{pmatrix}1,\ldots,n\\ 1,\ldots,n\end{pmatrix}}$$

where we use $\widehat{j}$ to indicate that we have deleted the jth index. Thus, in the numerator above, we have taken the determinant of the submatrix of A obtained by deleting the jth row and ith column. More generally we have

$$A^{-1}\begin{pmatrix}i_1,\ldots,i_p\\ j_1,\ldots,j_p\end{pmatrix} = \frac{(-1)^{\sum_{k=1}^p i_k+j_k} A\begin{pmatrix}j'_1,\ldots,j'_{n-p}\\ i'_1,\ldots,i'_{n-p}\end{pmatrix}}{A\begin{pmatrix}1,\ldots,n\\ 1,\ldots,n\end{pmatrix}}$$

where $i_1 < \cdots < i_p$ and $i'_1 < \cdots < i'_{n-p}$ are complementary indices in $\{1,\ldots,n\}$, as are the $j_1 < \cdots < j_p$ and $j'_1 < \cdots < j'_{n-p}$.

Laplace expansion by minors We will make use of the classical Laplace expansion of a determinant given by

$$A\begin{pmatrix}1,\ldots,n\\ 1,\ldots,n\end{pmatrix} = \sum_{1\le j_1<\cdots<j_p\le n} (-1)^{\sum_{r=1}^p i_r+j_r} A\begin{pmatrix}i_1,\ldots,i_p\\ j_1,\ldots,j_p\end{pmatrix} A\begin{pmatrix}i'_1,\ldots,i'_{n-p}\\ j'_1,\ldots,j'_{n-p}\end{pmatrix}.$$

In the above, $i_1 < \cdots < i_p$ and $i'_1 < \cdots < i'_{n-p}$ are complementary indices in $\{1,\ldots,n\}$, as are the $j_1 < \cdots < j_p$ and $j'_1 < \cdots < j'_{n-p}$; p is fixed; and the summation is over all ordered p-tuples $j_1 < \cdots < j_p$.

Principal minors If A is an $n \times n$ matrix, then its *principal submatrices* are those submatrices of the form

$$A\begin{bmatrix}i_1,\ldots,i_p\\ i_1,\ldots,i_p\end{bmatrix}.$$

That is to say, principal submatrices are the square submatrices of A, all of whose diagonal elements are diagonal elements of A. The *principal minors* of A are their determinants

$$A\begin{pmatrix} i_1, \ldots, i_p \\ i_1, \ldots, i_p \end{pmatrix}.$$

Determinantal identity The following determinantal identity will prove very useful. We state it here for easy reference. Let A be an $n \times m$ matrix. Let $1 \le i_1 < \cdots < i_r \le n$ and $1 \le j_1 < \cdots < j_{r+1} \le m$. Then for any $k \in \{1, \ldots, r\}$ and $\ell \in \{2, \ldots, r\}$

$$\begin{aligned} A&\begin{pmatrix} i_1, \ldots, i_r \\ j_1, \ldots, \widehat{j_\ell}, \ldots, j_{r+1} \end{pmatrix} A\begin{pmatrix} i_1, \ldots, \widehat{i_k}, \ldots, i_r \\ j_2, \ldots, j_r \end{pmatrix} \\ &= A\begin{pmatrix} i_1, \ldots, i_r \\ j_2, \ldots, j_{r+1} \end{pmatrix} A\begin{pmatrix} i_1, \ldots, \widehat{i_k}, \ldots, i_r \\ j_1, \ldots, \widehat{j_\ell}, \ldots, j_r \end{pmatrix} \\ &\quad + A\begin{pmatrix} i_1, \ldots, i_r \\ j_1, \ldots, j_r \end{pmatrix} A\begin{pmatrix} i_1, \ldots, \widehat{i_k}, \ldots, i_r \\ j_2, \ldots, \widehat{j_\ell}, \ldots, j_{r+1} \end{pmatrix}. \end{aligned} \tag{1.2}$$

Proof Let B be the $(r+1) \times (r+1)$ matrix given by

$$B = \begin{bmatrix} a_{i_1 j_1} & \cdots & a_{i_1 j_r} & a_{i_1 j_{r+1}} \\ \vdots & \ddots & \vdots & \vdots \\ a_{i_r j_1} & \cdots & a_{i_r j_r} & a_{i_r j_{r+1}} \\ 0 & \cdots & 0 & 1 \end{bmatrix}.$$

Apply Sylvester's Determinant Identity, where the pivot block of size $(r-1) \times (r-1)$ is given by all but the kth and last row, and all but the first and ℓth column of B. □

1.2 Building (strictly) totally positive matrices

If A is a totally positive or strictly totally positive matrix, then there are other (strictly) totally positive matrices associated with A and derived from A. There are also various operations that preserve the class of totally positive and strictly totally positive matrices. We review some of these here.

The first series of propositions as presented here are easily verified. Their proofs are left to the reader.

Proposition 1.2 *Assume A is a (strictly) totally positive matrix. Then A^T (the transpose of A), as well as every submatrix of A and A^T is (strictly) totally positive.*

Proposition 1.3 *Assume A is an $n \times m$ (strictly) totally positive matrix. Let B denote the matrix obtained from A by reversing the order of both its rows and columns, i.e., if $A = (a_{ij})_{i=1}^{n}{}_{j=1}^{m}$, then $B = (b_{ij})_{i=1}^{n}{}_{j=1}^{m}$, where $b_{ij} = a_{n+1-i,m+1-j}$, $i = 1, \ldots, n$, $j = 1, \ldots, m$. The matrix B is (strictly) totally positive.*

This next proposition immediately follows from an application of the Cauchy–Binet formula.

Proposition 1.4 *If A is an $n \times m$ totally positive matrix and B an $m \times r$ totally positive matrix, then AB is an $n \times r$ totally positive matrix. If $m \geq \min\{n, r\}$, A is an $n \times m$ strictly totally positive matrix and B an $m \times r$ totally positive matrix of rank r, then AB is an $n \times r$ strictly totally positive matrix. Similarly, if $m \geq \min\{n, r\}$, A is an $n \times m$ totally positive matrix of rank n and B an $m \times r$ strictly totally positive matrix, then AB is an $n \times r$ strictly totally positive matrix.*

Note that if $m < \min\{n, r\}$, then rank $AB \leq m$ and so AB cannot possibly be strictly totally positive.

Proposition 1.5 *The following operations preserve the class of (strictly) totally positive matrices.*

(i) *Multiplying a row (column) by a positive scalar.*
(ii) *Adding a positive multiple of a row (column) to the preceding or the succeeding row (column).*
(iii) *Adding a positive value to the $(1, 1)$ entry of the matrix (and to the (n, m) entry for an $n \times m$ matrix).*

From the formulæ for minors of the inverse we also have the following.

Proposition 1.6 *Assume A is a square strictly totally positive matrix. Then $DA^{-1}D$ is a strictly totally positive matrix, where D is the diagonal matrix with diagonal entries alternately 1 and -1. If A is a nonsingular totally positive matrix, then $DA^{-1}D$ is a nonsingular totally positive matrix.*

In addition, from Sylvester's Determinant Identity we have the following.

Proposition 1.7 *Assume A is an $n \times m$ (strictly) totally positive matrix. Fix $1 \le i_1 < \cdots < i_k \le n$ and $1 \le j_1 < \cdots < j_k \le m$, and let*

$$b_{ij} = A\begin{pmatrix} i_1, \ldots, i_k, i \\ j_1, \ldots, j_k, j \end{pmatrix}$$

for $i \in \{1, \ldots, n\} \backslash \{i_1, \ldots, i_k\}$ and $j \in \{1, \ldots, m\} \backslash \{j_1, \ldots, j_k\}$. (Recall that it is to be understood that we have arranged these row and column indices in natural order.) Then $B = (b_{ij})$ is an $(n-k) \times (m-k)$ (strictly) totally positive matrix.

A result that looks similar is the following:

Theorem 1.8 *Assume A is an $n \times m$ totally positive matrix. Given k, set*

$$b_{ij} = a_{ij} A\begin{pmatrix} 1, \ldots, k \\ 1, \ldots, k \end{pmatrix} - A\begin{pmatrix} 1, \ldots, k, i \\ 1, \ldots, k, j \end{pmatrix}$$

for $i = k+1, \ldots, n$, $j = k+1, \ldots, m$. Then $B = (b_{ij})$ is an $(n-k) \times (m-k)$ totally positive matrix. Furthermore, rank $B \le k$, and if A is a strictly totally positive matrix then all $r \times r$ minors of B are strictly positive for $r = 1, \ldots, \min\{k, n-k, m-k\}$.

We defer the proof of this theorem to the end of this chapter. There are additional operations that preserve strict total positivity and total positivity, but they are not as obvious or as immediate as some of the previous operations. We list two of them here, but defer their proof to Section 2.2.

Proposition 1.9 *Assume A is an $n \times m$ strictly totally positive matrix. For given $1 \le r < n$ set*

$$c_{ij} = \begin{cases} a_{ij}, & i = 1, \ldots, r, \ j = 2, \ldots, m \\ A\binom{i-1, i}{1, j}, & i = r+1, \ldots, n, \ j = 2, \ldots, m. \end{cases}$$

Then $C = (c_{ij})$ is an $n \times (m-1)$ strictly totally positive matrix. If A is a totally positive matrix, then C is a totally positive matrix.

Proposition 1.10 *Assume A is an $n \times m$ strictly totally positive matrix. Fix $p < \min\{m, n\}$ and set*

$$c_{ij} = A\begin{pmatrix} i-p, \ldots, i-1, i \\ 1, \ldots, p, j \end{pmatrix}$$

for $i = p+1, \ldots, n$, $j = p+1, \ldots, m$. Then $C = (c_{ij})$ is an $(n-p) \times (m-p)$ strictly totally positive matrix. If A is a totally positive matrix, then C is a totally positive matrix.

When $p = 1$, the above matrix C is a submatrix of the matrix C (with $r = 1$) in Proposition 1.9.

A totally different operation that preserves total positivity and strict total positivity is the following form of iteration. Let $A = (a_{ij})_{i,j=1}^n$ be an $n \times n$ matrix. We define the matrix $B = (b_{ij})_{i,j=1}^n$ as follows:

$$b_{1j} = a_{1j}, \qquad j = 1, \ldots, n,$$

and for $i \geq 2$

$$b_{ij} = \sum_{k=1}^{n} b_{i-1,k} a_{kj}, \qquad j = 1, \ldots, n.$$

Theorem 1.11 *Let A and B be as defined above. If A is a (strictly) totally positive matrix, then B is a (strictly) totally positive matrix.*

Proof We prove the theorem assuming A is strictly totally positive. The same proof holds, verbatim, for A totally positive.

Our proof will use induction arguments. From the definition we see that $b_{ij} > 0$ for all i, j. Assume that we have proven that all $p \times p$ minors of B are strictly positive for $p = 1, \ldots, r-1$. We prove that the same holds for all $r \times r$ minors.

Given $1 \leq i_1 < \cdots < i_r \leq n$, $1 \leq j_1 < \cdots < j_r \leq n$, consider

$$B\begin{pmatrix} i_1, \ldots, i_r \\ j_1, \ldots, j_r \end{pmatrix}.$$

We first assume that $i_1 = 1$. Expanding the above minor by its first row we obtain

$$B\begin{pmatrix} i_1, \ldots, i_r \\ j_1, \ldots, j_r \end{pmatrix} = \sum_{s=1}^{r} (-1)^{s-1} a_{1j_s} B\begin{pmatrix} i_2, \ldots, i_r \\ j_1, \ldots, \widehat{j_s}, \ldots, j_r \end{pmatrix}.$$

As $1 < i_2 < \cdots < i_r \leq n$, it follows from the Cauchy–Binet formula that

$$B\begin{pmatrix} i_2, \ldots, i_r \\ j_1, \ldots, \widehat{j_s}, \ldots, j_r \end{pmatrix} =$$
$$\sum_{1 \leq k_1 < \cdots < k_{r-1} \leq n} B\begin{pmatrix} i_2 - 1, \ldots, i_r - 1 \\ k_1, \ldots, k_{r-1} \end{pmatrix} A\begin{pmatrix} k_1, \ldots, k_{r-1} \\ j_1, \ldots, \widehat{j_s}, \ldots, j_r \end{pmatrix}.$$

Thus

$$B\begin{pmatrix} i_1,\dots,i_r \\ j_1,\dots,j_r \end{pmatrix} =$$

$$\sum_{s=1}^{r}(-1)^{s-1}a_{1j_s} \sum_{1\le k_1<\cdots<k_{r-1}\le n} B\begin{pmatrix} i_2-1,\dots,i_r-1 \\ k_1,\dots,k_{r-1} \end{pmatrix}$$

$$A\begin{pmatrix} k_1,\dots,k_{r-1} \\ j_1,\dots,\widehat{j_s},\dots,j_r \end{pmatrix}$$

$$= \sum_{1\le k_1<\cdots<k_{r-1}\le n} B\begin{pmatrix} i_2-1,\dots,i_r-1 \\ k_1,\dots,k_{r-1} \end{pmatrix} \sum_{s=1}^{r}(-1)^{s-1}a_{1j_s}$$

$$A\begin{pmatrix} k_1,\dots,k_{r-1} \\ j_1,\dots,\widehat{j_s},\dots,j_r \end{pmatrix}$$

$$= \sum_{2\le k_1<\cdots<k_{r-1}\le n} B\begin{pmatrix} i_2-1,\dots,i_r-1 \\ k_1,\dots,k_{r-1} \end{pmatrix} A\begin{pmatrix} 1,k_1,\dots,k_{r-1} \\ j_1,\dots,j_r \end{pmatrix}.$$

As each of the factors in the last sum is strictly positive (we use here the induction hypothesis) we have that

$$B\begin{pmatrix} i_1,\dots,i_r \\ j_1,\dots,j_r \end{pmatrix} > 0.$$

We complete the proof, for this fixed r, by applying an induction argument based on the value i_1. We have proved the result for $i_1 = 1$. Now assume that $i_1 > 1$. From the Cauchy–Binet formula,

$$B\begin{pmatrix} i_1,\dots,i_r \\ j_1,\dots,j_r \end{pmatrix} = \sum_{1\le k_1<\cdots<k_r\le n} B\begin{pmatrix} i_1-1,\dots,i_r-1 \\ k_1,\dots,k_r \end{pmatrix} A\begin{pmatrix} k_1,\dots,k_r \\ j_1,\dots,j_r \end{pmatrix}.$$

By our assumption on A and induction hypothesis on B each factor in the sum is strictly positive. This proves the theorem. □

This next result is interesting as we only vary columns and we only consider $n \times n$ minors of A (and thus do not really need the full total positivity of A to obtain our result).

Let A be an $n \times m$ matrix where $n < m$, and assume that

$$A\begin{pmatrix} 1,\dots,n \\ 1,\dots,n \end{pmatrix} \neq 0.$$

We define the $n \times (m-n)$ matrix $B = (b_{ij})$ as follows:

$$b_{ij} = A\begin{pmatrix} 1,\dots,n \\ 1,\dots,\widehat{i},\dots,n,n+j \end{pmatrix}, \qquad i = 1,\dots,n,\ j = 1,\dots,m-n.$$

Then,

Theorem 1.12 *For A and B as defined above,*

$$B\binom{i_1,\ldots,i_r}{j_1,\ldots,j_r} = (-1)^{\frac{r(r-1)}{2}}\left[A\binom{1,\ldots,n}{1,\ldots,n}\right]^{r-1} A\binom{1,\ldots,n}{i'_1,\ldots,i'_{n-r},n+j_1,\ldots,n+j_r}$$

where $i'_1,\ldots,i'_{n-r}$ is complementary to $i_1,\ldots,i_r$ in $\{1,\ldots,n\}$. Set $C = (c_{ij})_{i=1}^{n}{}_{j=1}^{m-n}$ where

$$c_{ij} = b_{n-i+1,j}, \qquad i=1,\ldots,n,\ j=1,\ldots,m-n.$$

If A is a (strictly) totally positive matrix, then C is a (strictly) totally positive matrix.

Proof Let D be the $2n\times m$ matrix whose first n rows are the unit vectors $\mathbf{e}^i$, $i=1,\ldots,n$, and whose last n rows are A. We apply Sylvester's Determinant Identity with pivot block

$$D\begin{bmatrix} n+1,\ldots,2n \\ 1,\ldots,n\end{bmatrix}.$$

That is, set

$$e_{ij} = D\binom{i,n+1,\ldots,2n}{1,\ldots,n,n+j}, \qquad i=1,\ldots,n,\ j=1,\ldots,m-n.$$

Note that

$$e_{ij} = (-1)^{i+1}A\binom{1,\ldots,n}{1,\ldots,\widehat{i},\ldots,n,n+j} = (-1)^{i+1}b_{ij}.$$

Therefore, from Sylvester's Determinant Identity,

$$\begin{aligned} B\binom{i_1,\ldots,i_r}{j_1,\ldots,j_r} &= (-1)^{r+\sum_{k=1}^r i_k} E\binom{i_1,\ldots,i_r}{j_1,\ldots,j_r} \\ &= (-1)^{r+\sum_{k=1}^r i_k}\left[D\binom{n+1,\ldots,2n}{1,\ldots,n}\right]^{r-1} \\ &\qquad D\binom{i_1,\ldots,i_r,n+1,\ldots,2n}{1,\ldots,n,n+j_1,\ldots,n+j_r}. \end{aligned}$$

Now

$$D\binom{n+1,\ldots,2n}{1,\ldots,n} = A\binom{1,\ldots,n}{1,\ldots,n}$$

while

$$D\begin{pmatrix} i_1,\ldots,i_r,n+1,\ldots,2n \\ 1,\ldots,n,n+j_1,\ldots,n+j_r \end{pmatrix} =$$
$$(-1)^{\sum_{k=1}^{r} i_k+k} A\begin{pmatrix} 1,\ldots,n \\ i'_1,\ldots,i'_{n-r},n+j_1,\ldots,n+j_r \end{pmatrix}$$

where $i'_1,\ldots,i'_{n-r}$ is complementary to $i_1,\ldots,i_r$ in $\{1,\ldots,n\}$.

Thus

$$B\begin{pmatrix} i_1,\ldots,i_r \\ j_1,\ldots,j_r \end{pmatrix} =$$
$$(-1)^{\frac{r(r-1)}{2}} \left[A\begin{pmatrix} 1,\ldots,n \\ 1,\ldots,n \end{pmatrix}\right]^{r-1} A\begin{pmatrix} 1,\ldots,n \\ i'_1,\ldots,i'_{n-r},n+j_1,\ldots,n+j_r \end{pmatrix}.$$

The matrix C is obtained from B by simply interchanging the order of the rows and therefore

$$C\begin{pmatrix} i_1,\ldots,i_r \\ j_1,\ldots,j_r \end{pmatrix} = \left[A\begin{pmatrix} 1,\ldots,n \\ 1,\ldots,n \end{pmatrix}\right]^{r-1} A\begin{pmatrix} 1,\ldots,n \\ i'_1,\ldots,i'_{n-r},n+j_1,\ldots,n+j_r \end{pmatrix}.$$

The result now follows. □

There is another construction of strictly totally positive matrices, related to the previous construction. Assume A is an $n \times m$ strictly totally positive matrix. Form the $(n+m) \times m$ matrix D by adjoining to A the $m \times m$ matrix C given by

$$C = \begin{pmatrix} 0 & 0 & \cdots & 0 & 1 \\ 0 & 0 & \cdots & -1 & 0 \\ \vdots & \vdots & \ddots & \vdots & \vdots \\ (-1)^{m-1} & 0 & \cdots & 0 & 0 \end{pmatrix},$$

i.e.,

$$D = \begin{bmatrix} A \\ C \end{bmatrix}.$$

It is readily verified that

$$D\begin{pmatrix} \alpha_1,\ldots,\alpha_m \\ 1,\ldots,m \end{pmatrix} = A\begin{pmatrix} i_1,\ldots,i_k \\ j_1,\ldots,j_k \end{pmatrix}$$

where $\{\alpha_1,\ldots,\alpha_m\} = \{i_1,\ldots,i_k,n+m+1-j'_{m-k},\ldots,n+m+1-j'_1\}$, and $\{j'_1,\ldots,j'_{m-k}\}$ is complementary to $\{j_1,\ldots,j_k\}$ in $\{1,\ldots,m\}$. When $k=0$, i.e., $(\alpha_1,\ldots,\alpha_m) = (n+1,\ldots,n+m)$, we set $A(\emptyset) = 1$ in agreement with the above equality. Thus the strict total positivity property of A is equivalent to the fact

$$D\begin{pmatrix}\alpha_1,\ldots,\alpha_m\\1,\ldots,m\end{pmatrix} > 0$$

for all $1 \le \alpha_1 < \cdots < \alpha_m \le n+m$.

Two elementary operations that preserve this latter property of D are the following. First, we can cyclically shift the rows of D, where we multiply the row going from the first to the last (or last to first) by $(-1)^{m-1}$. Second, we can multiply D by any $m \times m$ matrix M with $\det M > 0$.

Let E be the $(n+m) \times m$ matrix obtained from D by a simple forward cyclic rotation of the rows, i.e., shift row r to row $r+1$ and row $n+m$ to row 1, and multiply the new first row of E by $(-1)^{m-1}$. In addition, since

$$E\begin{pmatrix}n+1,\ldots,n+m\\1,\ldots,m\end{pmatrix} = D\begin{pmatrix}n,n+1,\ldots,n+m-1\\1,\ldots,m\end{pmatrix} > 0$$

there exists an $m \times m$ matrix M with $\det M > 0$ such that

$$F = EM = \begin{bmatrix}B\\C\end{bmatrix}$$

where C is as was previously defined. Thus B is an $n \times m$ strictly totally positive matrix. A calculation shows that

$$B = \begin{pmatrix} a_{n,2} & a_{n,3} & \cdots & a_{n,m} & 1\\ A\binom{1,n}{1,2} & A\binom{1,n}{1,3} & \cdots & A\binom{1,n}{1,m} & a_{1,1}\\ \vdots & \vdots & \ddots & \vdots & \vdots\\ A\binom{n-1,n}{1,2} & A\binom{n-1,n}{1,3} & \cdots & A\binom{n-1,n}{1,m} & a_{n-1,1}\end{pmatrix}.$$

It is also easily verified by direct calculation that this matrix is strictly totally positive. In the above it is not sufficient that A be totally positive.

1.3 Nonsingularity and rank

Totally positive nonsingular matrices enjoy surprising and useful properties, one of which is the following:

Theorem 1.13 *Let A be an $n \times n$ totally positive nonsingular matrix. Then all principal minors of A are strictly positive.*

Prior to proving this result we first explain and prove an ancillary result that will be of independent interest.

Zero entries of totally positive matrices and zero values of their minors are evidence of boundary behavior within the class of totally positive matrices and, as such, are not arbitrary in nature. A zero entry of a totally positive matrix A or a zero minor of this totally positive matrix portends linear dependence or "throws a shadow." That is, under suitable linear independence assumptions all minors of the same order to the right and above it, or to the left and below it, are also zero. Let us define these notions more precisely.

Definition 1.14 Let $A = (a_{ij})_{i=1}^{n}{}_{j=1}^{m}$.

(a) The *right shadow* of the entry a_{ij} is the $i \times (m - j + 1)$ submatrix $(a_{rs})_{r=1}^{i}{}_{s=j}^{m}$.

(b) The *left shadow* of the entry a_{ij} is the $(n - i + 1) \times j$ submatrix $(a_{rs})_{r=i}^{n}{}_{s=1}^{j}$.

(c) The *right shadow* of the submatrix

$$A\begin{bmatrix} i+1,\ldots,i+r \\ j+1,\ldots,j+r \end{bmatrix}$$

is the $(i + r) \times (m - j)$ submatrix

$$A\begin{bmatrix} 1,\ldots,i+r \\ j+1,\ldots,m \end{bmatrix}.$$

(d) The *left shadow* of the submatrix

$$A\begin{bmatrix} i+1,\ldots,i+r \\ j+1,\ldots,j+r \end{bmatrix}$$

is the $(n - i) \times (j + r)$ submatrix

$$A\begin{bmatrix} i+1,\ldots,n \\ 1,\ldots,j+r \end{bmatrix}.$$

Note that in this definition we only consider submatrices with consecutive rows and columns. This is not without reason, as will become evident from Theorem 1.19. We can now state and prove:

Proposition 1.15 *If A is an $n \times m$ totally positive matrix and*

$$\operatorname{rank} A\begin{bmatrix} i+1,\ldots,i+r \\ j+1,\ldots,j+r \end{bmatrix} = r - 1$$

then at least one of the following holds. Either the rows $i + 1, \ldots, i + r$ or the columns $j + 1, \ldots, j + r$ of A are linearly dependent, or the right or left *shadow of*

$$A\begin{bmatrix} i+1,\ldots,i+r \\ j+1,\ldots,j+r \end{bmatrix}$$

has rank $r-1$.

Proof We first prove the case $r=1$. That is, we prove that if A is totally positive and $a_{ij}=0$, then at least one of the following holds. Either the ith row or jth column of A is zero, or the right or left shadow of a_{ij} is zero.

Assume neither the ith row nor the jth column of A is zero. Let $a_{i\ell}>0$ for some ℓ. If $\ell<j$ we prove that the right shadow of A is zero as follows. For any $r<i$

$$0 \le A\begin{pmatrix} r,i \\ \ell,j \end{pmatrix} = a_{r\ell}a_{ij} - a_{rj}a_{i\ell} = -a_{rj}a_{i\ell} \le 0.$$

Since $a_{i\ell}>0$ we have $a_{rj}=0$ for all $r<i$. As the jth column of A is not zero we have a $k>i$ for which $a_{kj}>0$. Now, for any $r\le i$, $s\ge j$,

$$0 \le A\begin{pmatrix} r,k \\ j,s \end{pmatrix} = a_{rj}a_{ks} - a_{rs}a_{kj} = -a_{rs}a_{kj} \le 0,$$

and since $a_{kj}>0$ we have $a_{rs}=0$. Similarly, if $\ell>j$ then it follows that the left shadow of a_{ij} is zero.

Let us now assume that $r>1$. As

$$A\begin{bmatrix} i+1,\ldots,i+r \\ j+1,\ldots,j+r \end{bmatrix}$$

is of rank $r-1$ there are $p,q\in\{1,\ldots,r\}$ such that

$$A\begin{pmatrix} i+1,\ldots,\widehat{i+p},\ldots,i+r \\ j+1,\ldots,\widehat{j+q},\ldots,j+r \end{pmatrix} > 0. \tag{1.3}$$

Set

$$b_{k\ell} = A\begin{pmatrix} i+1,\ldots,\widehat{i+p},\ldots,i+r,k \\ j+1,\ldots,\widehat{j+q},\ldots,j+r,\ell \end{pmatrix}$$

for $k\in\{1,\ldots,n\}\backslash\{i+1,\ldots,\widehat{i+p},\ldots,i+r\}$ and $\ell\in\{1,\ldots,m\}\backslash\{j+1,\ldots,\widehat{j+p},\ldots,j+r\}$, where we understand that the row and column indices are rearranged in increasing order. Let $B=(b_{k\ell})$. By Sylvester's Determinant Identity, we see that B is an $(n-r+1)\times(m-r+1)$ totally positive matrix and $b_{i+p,j+q}=0$. As such we can apply the result proved in the case $r=1$.

If $b_{i+p,\ell} = 0$ for all $\ell \in \{1,\ldots,m\}\backslash\{j+1,\ldots,\widehat{j+q},\ldots,j+r\}$, i.e.,

$$A\begin{pmatrix} i+1,\ldots,i+r \\ j+1,\ldots,\widehat{j+q},\ldots,j+r,\ell \end{pmatrix} = 0$$

for all ℓ, then from (1.3) it follows that the rows $i+1,\ldots,i+r$ are linearly dependent. Similarly, if $b_{k,j+q} = 0$ for all $k \in \{1,\ldots,n\}\backslash\{i+1,\ldots,\widehat{i+p},\ldots,i+r\}$, then the columns $j+1,\ldots,j+r$ are linearly dependent.

If the right shadow of $b_{i+p,j+q}$ vanishes, i.e.,

$$A\begin{pmatrix} i+1,\ldots,\widehat{i+p},\ldots,i+r,k \\ j+1,\ldots,\widehat{j+q},\ldots,j+r,\ell \end{pmatrix} = 0$$

for all $k \le i+p$, $\ell \ge j+q$, then from (1.3) we see that

$$A\begin{bmatrix} 1,\ldots,i+r \\ j+1,\ldots,m \end{bmatrix}$$

is of rank $r-1$. Similarly, if the left shadow of $b_{i+p,j+q}$ vanishes, then

$$A\begin{bmatrix} i+1,\ldots,n \\ 1,\ldots,j+r \end{bmatrix}$$

is of rank $r-1$. □

Proposition 1.15, in the case $r = 1$, is sometimes referred to as the "shadow lemma." We now prove Theorem 1.13.

Proof of Theorem 1.13 We first prove directly that $a_{rr} > 0$ for all $r \in \{1,\ldots,n\}$. Assume $a_{rr} = 0$. From Proposition 1.15 we have four options. But all four options contradict the nonsingularity of A. Obviously we cannot have that the rth row or column of A is zero. Thus either the left or right shadow of a_{rr} is zero. Assume it is the right shadow. Then $a_{ij} = 0$ for all $i \le r$ and all $j \ge r$, implying that the first r rows of A are linearly dependent. This is a contradiction and therefore $a_{rr} > 0$.

We derive the general result by applying an induction argument on the size of the minor and using Sylvester's Determinant Identity. We assume that for any totally positive nonsingular $n \times n$ matrix (any n) all principal minors of order at most $p-1$ are strictly positive ($p \le n$). We prove that this same result holds for all principal minors of order p . We have proved the case $p = 1$. For any $1 \le i_1 < \cdots < i_p \le n$ set

$$b_{k\ell} = A\begin{pmatrix} i_1,\ldots,i_{p-1},k \\ i_1,\ldots,i_{p-1},\ell \end{pmatrix},$$

for $k, \ell \in \{1, \ldots, n\} \backslash \{i_1, \ldots, i_{p-1}\}$, and let $B = (b_{k\ell})$. As an immediate consequence of Sylvester's Determinant Identity and our induction hypothesis, it follows that B is totally positive and nonsingular. Thus the diagonal entries of B are strictly positive. As

$$0 < b_{i_p i_p} = \left[A\begin{pmatrix} i_1, \ldots, i_{p-1} \\ i_1, \ldots, i_{p-1} \end{pmatrix}\right]^{p-2} A\begin{pmatrix} i_1, \ldots, i_p \\ i_1, \ldots, i_p \end{pmatrix}$$

and, by our induction hypothesis, we have

$$A\begin{pmatrix} i_1, \ldots, i_{p-1} \\ i_1, \ldots, i_{p-1} \end{pmatrix} > 0$$

it therefore follows that

$$A\begin{pmatrix} i_1, \ldots, i_p \\ i_1, \ldots, i_p \end{pmatrix} > 0.$$

□

For totally positive matrices there is also an interaction between its rank, which of its rows and columns can be linearly (in)dependent, and the strict positivity of specific minors. This we detail in the next series of results.

Proposition 1.16 *Let A be an $n \times m$ totally positive matrix, and let $\mathbf{a}^k$ denote the kth row of A, $k = 1, \ldots, n$. Given $1 = i_1 < \cdots < i_{r+1} = n$, assume that the $r+1$ vectors $\mathbf{a}^{i_1}, \ldots, \mathbf{a}^{i_{r+1}}$ are linearly dependent, while the r vectors $\mathbf{a}^{i_1}, \ldots, \mathbf{a}^{i_r}$ and $\mathbf{a}^{i_2}, \ldots, \mathbf{a}^{i_{r+1}}$ are each linearly independent. Then A is necessarily of rank r.*

Proof Since the $\mathbf{a}^{i_1}, \ldots, \mathbf{a}^{i_{r+1}}$ are linearly dependent, while the $\mathbf{a}^{i_2}, \ldots, \mathbf{a}^{i_{r+1}}$ are linearly independent we can write

$$\mathbf{a}^{i_1} = \sum_{s=2}^{r+1} c_s \mathbf{a}^{i_s}.$$

As the $\mathbf{a}^{i_1}, \ldots, \mathbf{a}^{i_r}$ are linearly independent there exist $1 \le j_1 < \cdots < j_r \le m$ for which

$$A\begin{pmatrix} i_1, \ldots, i_r \\ j_1, \ldots, j_r \end{pmatrix} > 0.$$

Substituting for $\mathbf{a}^{i_1}$ we obtain

$$\begin{aligned} A\begin{pmatrix} i_1, \ldots, i_r \\ j_1, \ldots, j_r \end{pmatrix} &= \sum_{s=2}^{r+1} c_s A\begin{pmatrix} i_s, i_2, \ldots, i_r \\ j_1, j_2, \ldots, j_r \end{pmatrix} = c_{r+1} A\begin{pmatrix} i_{r+1}, i_2, \ldots, i_r \\ j_1, j_2, \ldots, j_r \end{pmatrix} \\ &= (-1)^{r+1} c_{r+1} A\begin{pmatrix} i_2, \ldots, i_r, i_{r+1} \\ j_1, \ldots, j_{r-1}, j_r \end{pmatrix}. \end{aligned}$$

The matrix A is totally positive and the left-hand side is strictly positive. Thus $(-1)^{r+1}c_{r+1} > 0$.

By assumption A is of rank at least r. If $r = m$ there is nothing to prove, while if $r + 1 = n$ there is also nothing to prove. As such we assume that $r + 1 \le m$ and $r + 1 < n$. Let $\ell \in \{1, \ldots, n\} \backslash \{i_1, \ldots, i_{r+1}\}$. Thus $1 = i_1 < \ell < i_{r+1} = n$. For every choice of $1 \le k_1 < \cdots < k_{r+1} \le m$

$$\begin{aligned} A\begin{pmatrix} i_1, \ldots, \ell, \ldots, i_r \\ k_1, k_2, \ldots, k_{r+1} \end{pmatrix} &= \sum_{s=2}^{r+1} c_s A\begin{pmatrix} i_s, i_2, \ldots, \ell, \ldots, i_r \\ k_1, k_2, \ldots, k_{r+1} \end{pmatrix} \\ &= c_{r+1} A\begin{pmatrix} i_{r+1}, i_2, \ldots, \ell, \ldots, i_r \\ k_1, k_2, \ldots, k_{r+1} \end{pmatrix} \\ &= (-1)^r c_{r+1} A\begin{pmatrix} i_2, \ldots, \ell, \ldots, i_r, i_{r+1} \\ k_1, k_2, \ldots, k_{r+1} \end{pmatrix}. \end{aligned}$$

As A is totally positive and $(-1)^r c_{r+1} < 0$, it follows that

$$A\begin{pmatrix} i_1, \ldots, \ell, \ldots, i_r \\ k_1, k_2, \ldots, k_{r+1} \end{pmatrix} = 0.$$

Since this is true for every choice of $1 \le k_1 < \cdots < k_{r+1} \le m$, we have that $\mathbf{a}^\ell \in \operatorname{span}\{\mathbf{a}^{i_1}, \ldots, \mathbf{a}^{i_r}\}$. From our assumption we also have $\mathbf{a}^{i_{r+1}} \in \operatorname{span}\{\mathbf{a}^{i_1}, \ldots, \mathbf{a}^{i_r}\}$. Thus A is of rank r. □

Proposition 1.17 *Let A be an $n \times m$ totally positive matrix. Assume $1 = i_1 < \cdots < i_{r+1} = n$ and $1 = j_1 < \cdots < j_{r+1} = m$. If*

$$A\begin{pmatrix} i_1, \ldots, i_{r+1} \\ j_1, \ldots, j_{r+1} \end{pmatrix} = 0,$$

while

$$A\begin{pmatrix} i_1, \ldots, i_r \\ j_1, \ldots, j_r \end{pmatrix}, \; A\begin{pmatrix} i_2, \ldots, i_{r+1} \\ j_2, \ldots, j_{r+1} \end{pmatrix} > 0,$$

then A is of rank r.

Proof Let $\mathbf{a}^k$ denote the kth row of A. By assumption the r vectors $\mathbf{a}^{i_1}, \ldots, \mathbf{a}^{i_r}$ and the r vectors $\mathbf{a}^{i_2}, \ldots, \mathbf{a}^{i_{r+1}}$ are linearly independent. Thus it suffices, by Proposition 1.16, to prove that the $r+1$ vectors $\mathbf{a}^{i_1}, \ldots, \mathbf{a}^{i_{r+1}}$ are linearly dependent.

This latter result follows from an application of Proposition 1.16 to the $(r+1) \times m$ matrix

$$B = A\begin{bmatrix} i_1, \ldots, i_{r+1} \\ 1, \ldots, m \end{bmatrix}$$

where we exchange the roles of the rows and columns. That is, if $\mathbf{b}^k$ denotes the kth column of B, then, by assumption, the $r+1$ vectors $\mathbf{b}^{j_1}, \ldots, \mathbf{b}^{j_{r+1}}$ are linearly dependent, while the r vectors $\mathbf{b}^{j_1}, \ldots, \mathbf{b}^{j_r}$ and $\mathbf{b}^{j_2}, \ldots, \mathbf{b}^{j_{r+1}}$ are each linearly independent. Thus B is of rank r, implying that the $r+1$ vectors $\mathbf{a}^{i_1}, \ldots, \mathbf{a}^{i_{r+1}}$ are linearly dependent. □

Another consequence of Proposition 1.16 is the following, which tells us something about the possible singularity structure of totally positive matrices.

Proposition 1.18 *Let A be an $n \times m$ totally positive matrix, and let $\mathbf{a}^k$ denote the kth row of A, $k = 1, \ldots, n$. Given $1 = i_1 < \cdots < i_{r+1} < n$, assume that the $r+1$ vectors $\mathbf{a}^{i_1}, \ldots, \mathbf{a}^{i_{r+1}}$ are linearly independent. Let $k > i_{r+1}$ ($k \le n$) and assume that the $r+1$ vectors $\mathbf{a}^{i_1}, \ldots, \mathbf{a}^{i_r}, \mathbf{a}^k$ are linearly dependent. Then $\mathbf{a}^k = \mathbf{0}$.*

Proof This proposition can be proved using the method of proof in Proposition 1.16. But it also follows as a consequence of Proposition 1.16. We explain this latter proof here.

We have that the rows $\mathbf{a}^{i_1}, \ldots, \mathbf{a}^{i_r}, \mathbf{a}^k$ are linearly dependent, while the $\mathbf{a}^{i_1}, \ldots, \mathbf{a}^{i_r}$ are linearly independent. If the rows $\mathbf{a}^{i_2}, \ldots, \mathbf{a}^{i_r}, \mathbf{a}^k$ are linearly independent, then, from Proposition 1.16, the rank of the matrix composed from the rows indexed $i_1, \ldots, i_r, i_{r+1}, k$ is exactly r. But, by assumption, the $\mathbf{a}^{i_1}, \ldots, \mathbf{a}^{i_{r+1}}$ are linearly independent. Thus the $\mathbf{a}^{i_2}, \ldots, \mathbf{a}^{i_r}, \mathbf{a}^k$ are necessarily linearly dependent. We repeat this argument, each time lopping off the first vector, until we arrive at the desired result that $\mathbf{a}^k$ is linearly dependent, i.e., $\mathbf{a}^k$ is the zero vector. □

The above results allow us to totally characterize the possible vanishing minors of a nonsingular totally positive matrix. We have the following.

Theorem 1.19 *Let A be an $n \times n$ nonsingular totally positive matrix. Let $(\alpha_k, \beta_k, r_k)_{k=1}^{\ell}$ be the set of all triples such that for each $k \in \{1, \ldots, \ell\}$*

$$A\begin{pmatrix} \alpha_k + 1, \ldots, \alpha_k + r_k \\ \beta_k + 1, \ldots, \beta_k + r_k \end{pmatrix} = 0,$$

and no principal minors of the $r_k \times r_k$ submatrix

$$A\begin{bmatrix} \alpha_k + 1, \ldots, \alpha_k \mid r_k \\ \beta_k + 1, \ldots, \beta_k + r_k \end{bmatrix}$$

vanish. For such submatrices we have $\alpha_k \neq \beta_k$*. Moreover*

$$A\begin{pmatrix} i_1,\ldots,i_p \\ j_1,\ldots,j_p \end{pmatrix} = 0$$

if and only if for some $k \in \{1,\ldots,\ell\}$ *there is an* $r_k \times r_k$ *principal submatrix of*

$$A\begin{bmatrix} i_1,\ldots,i_p \\ j_1,\ldots,j_p \end{bmatrix}$$

that lies in the right shadow of an above

$$A\begin{bmatrix} \alpha_k+1,\ldots,\alpha_k+r_k \\ \beta_k+1,\ldots,\beta_k+r_k \end{bmatrix}$$

if $\alpha_k < \beta_k$*, or in its left shadow if* $\alpha_k > \beta_k$*.*

In other words, what we are proving here is that for a nonsingular totally positive matrix there are certain basic zero minors formed from consecutive rows and columns and all other zero minors are derived from them. This derivation is also of a very specific nature. It is a consequence of a vanishing principal minor of the zero minor lying in the shadow of one of these basic zero minors.

Proof Assume

$$A\begin{pmatrix} \alpha_k+1,\ldots,\alpha_k+r_k \\ \beta_k+1,\ldots,\beta_k+r_k \end{pmatrix} = 0,$$

and no principal minor of

$$A\begin{bmatrix} \alpha_k+1,\ldots,\alpha_k+r_k \\ \beta_k+1,\ldots,\beta_k+r_k \end{bmatrix}$$

vanishes.

As A is nonsingular it follows from Theorem 1.13 that $\alpha_k \neq \beta_k$. In addition, from Proposition 1.15 we have that each such vanishing minor either throws a right or a left shadow. If $\alpha_k < \beta_k$ then it must throw a right shadow since the left shadow of

$$A\begin{bmatrix} \alpha_k+1,\ldots,\alpha_k+r_k \\ \beta_k+1,\ldots,\beta_k+r_k \end{bmatrix}$$

is

$$A\begin{bmatrix} \alpha_k+1,\ldots,n \\ 1,\ldots,\beta_k+r_k \end{bmatrix}$$

which contains the nonsingular $r_k \times r_k$ principal submatrix

$$A\begin{bmatrix} \alpha_k+1,\ldots,\alpha_k+r_k \\ \alpha_k+1,\ldots,\alpha_k+r_k \end{bmatrix}.$$

But this contradicts Theorem 1.13. Similarly if $\alpha_k > \beta_k$ then

$$A\begin{bmatrix}\alpha_k+1,\ldots,\alpha_k+r_k\\ \beta_k+1,\ldots,\beta_k+r_k\end{bmatrix}$$

must throw a left shadow.

Now if

$$A\begin{bmatrix}i_1,\ldots,i_p\\ j_1,\ldots,j_p\end{bmatrix}$$

is such that it contains an $r_k \times r_k$ principal submatrix that lies in the right shadow of one of the

$$A\begin{bmatrix}\alpha_k+1,\ldots,\alpha_k+r_k\\ \beta_k+1,\ldots,\beta_k+r_k\end{bmatrix}$$

if $\alpha_k < \beta_k$, or in its left shadow if $\alpha_k > \beta_k$, then from Proposition 1.15 this principal minor of

$$A\begin{bmatrix}i_1,\ldots,i_p\\ j_1,\ldots,j_p\end{bmatrix}$$

must vanish. It now follows from Theorem 1.13 that we must have

$$A\begin{pmatrix}i_1,\ldots,i_p\\ j_1,\ldots,j_p\end{pmatrix}=0.$$

This proves the easier direction of the theorem.

Let us now assume that

$$A\begin{pmatrix}i_1,\ldots,i_p\\ j_1,\ldots,j_p\end{pmatrix}=0$$

for some choice of $1 \le i_1 < \cdots < i_p \le n$ and $1 \le j_1 < \cdots < j_p \le m$. We assume, in what follows, that no principal minors of this minor vanish. (Otherwise replace the above by a principal minor with this same property.) As

$$A\begin{pmatrix}i_1,\ldots,i_p\\ j_1,\ldots,j_p\end{pmatrix}=0$$

while

$$A\begin{pmatrix}i_1,\ldots,i_{p-1}\\ j_1,\ldots,j_{p-1}\end{pmatrix},\ A\begin{pmatrix}i_2,\ldots,i_p\\ j_2,\ldots,j_p\end{pmatrix}>0$$

it follows from Proposition 1.17 that the $(i_p-i_1+1)\times(j_p-j_1+1)$ totally positive matrix

$$A\begin{bmatrix}i_1,i_1+1,\ldots,i_p\\ j_1,j_1+1,\ldots,j_p\end{bmatrix}$$

(composed of consecutive rows and columns) has rank $p-1$. In this proof we can assume $p > 1$ as the case $p = 1$ is a direct consequence of Proposition 1.15.

We claim that

$$j_p - i_1 \le p - 2 \qquad \text{or} \qquad i_p - j_1 \le p - 2.$$

Assume not. Then

$$j_p - i_1 \ge p - 1 \qquad \text{and} \qquad i_p - j_1 \ge p - 1. \tag{1.4}$$

Set

$$\alpha = \max\{i_1, j_1\} - 1$$

We claim that

$$i_1 \le \alpha + 1 < \cdots < \alpha + p \le i_p$$

and

$$j_1 \le \alpha + 1 < \cdots < \alpha + p \le j_p.$$

To see this note that if $\alpha = i_1 - 1$, then $\alpha + p = i_1 + p - 1 \le i_p$ (since the $i_1, \ldots, i_p$ are increasing integers), while $\alpha + 1 = i_1 \ge j_1$ and from (1.4) $\alpha + p = i_1 + p - 1 \le j_p$. If $\alpha = j_1 - 1$ the same analysis applies. As

$$A\begin{bmatrix} i_1, i_1 + 1, \ldots, i_p \\ j_1, j_1 + 1, \ldots, j_p \end{bmatrix}$$

has rank $p - 1$, it therefore follows that

$$A\begin{pmatrix} \alpha + 1, \ldots, \alpha + p \\ \alpha + 1, \ldots, \alpha + p \end{pmatrix} = 0,$$

which contradicts Theorem 1.13. Thus

$$j_p - i_1 \le p - 2 \qquad \text{or} \qquad i_p - j_1 \le p - 2.$$

Let us assume that $j_p - i_1 \le p - 2$. The matrix

$$A\begin{bmatrix} i_1, i_1 + 1, \ldots, i_1 - p + 1 \\ j_p - p + 1, j_p - p + 2, \ldots, j_p \end{bmatrix}$$

composed of p consecutive rows and columns is of rank, at most, $p-1$. Thus by Proposition 1.15 this submatrix throws a right or left shadow. From the analysis of the first part of the proof of this theorem we see that it throws a left shadow since $i_1 > j_p - p + 1$. That is,

$$A\begin{bmatrix} i_1, \ldots, n \\ 1, \ldots, j_p \end{bmatrix}$$

is of rank $p-1$. As

$$A\begin{bmatrix} i_1,\dots,i_p \\ j_1,\dots,j_p \end{bmatrix}$$

lies in this submatrix we have that

$$A\begin{bmatrix} i_1,\dots,i_p \\ j_1,\dots,j_p \end{bmatrix}$$

lies in the left shadow of

$$A\begin{bmatrix} \alpha+1,\dots,\alpha+p \\ \beta+1,\dots,\beta+p \end{bmatrix}$$

where $\alpha = i_1 - 1$ and $\beta = j_p - p$. In fact from our assumption that no principal minors of

$$A\begin{bmatrix} i_1,\dots,i_p \\ j_1,\dots,j_p \end{bmatrix}$$

vanish, it necessarily follows that

$$A\begin{bmatrix} \alpha+1,\dots,\alpha+p \\ \beta+1,\dots,\beta+p \end{bmatrix}$$

is of rank exactly $p-1$.

The case where $i_p - j_1 \le p-2$ is handled similarly. That is, it follows that the

$$A\begin{bmatrix} i_1,\dots,i_p \\ j_1,\dots,j_p \end{bmatrix}$$

is in the right shadow of the matrix

$$A\begin{bmatrix} i_p-p+1, i_p-p+2,\dots,i_p \\ j_1, j_1+1,\dots,j_1+p-1 \end{bmatrix}$$

of rank $p-1$. This proves the theorem. □

One simple application of Theorem 1.19 is the following. As we have seen from Theorem 1.13, if A is totally positive and

$$A\begin{pmatrix} i_1,\dots,i_p \\ j_1,\dots,j_p \end{pmatrix} > 0$$

then $a_{i_k,j_k} > 0$, $k = 1,\dots,p$. In general the converse need not hold. We claim that if the converse holds for consecutive row and column indices, then it holds in general.

Proposition 1.20 *Let A be an $n \times n$ nonsingular totally positive matrix. Assume that*

$$A\begin{pmatrix} i+1,\ldots,i+p \\ j+1,\ldots,j+p \end{pmatrix} > 0$$

if $a_{i+k,j+k} > 0$, $k = 1,\ldots,p$, for all possible i, j and p. Then for all $1 \le i_1 < \cdots < i_p \le n$, $1 \le j_1 < \cdots < j_p \le n$, and all p we have

$$A\begin{pmatrix} i_1,\ldots,i_p \\ j_1,\ldots,j_p \end{pmatrix} > 0$$

if $a_{i_k,j_k} > 0$, $k = 1,\ldots,p$.

Proof Assume

$$A\begin{pmatrix} i_1,\ldots,i_p \\ j_1,\ldots,j_p \end{pmatrix} = 0.$$

From Theorem 1.19 there exist (α,β,r) such that

$$A\begin{pmatrix} \alpha+1,\ldots,\alpha+r \\ \beta+1,\ldots,\beta+r \end{pmatrix} = 0,$$

no principal submatrix of

$$A\begin{bmatrix} \alpha+1,\ldots,\alpha+r \\ \beta+1,\ldots,\beta+r \end{bmatrix}$$

vanishes, and some $r \times r$ principal submatrix of

$$A\begin{bmatrix} i_1,\ldots,i_p \\ j_1,\ldots,j_p \end{bmatrix}$$

lies in the right shadow of

$$A\begin{bmatrix} \alpha+1,\ldots,\alpha+r \\ \beta+1,\ldots,\beta+r \end{bmatrix}$$

if $\alpha < \beta$, or in its left shadow if $\alpha > \beta$.

The assumption of the proposition implies that $r = 1$ and thus $a_{\alpha+1,\beta+1} = 0$. We therefore have an $\alpha \neq \beta$ such that for some $k \in \{1,\ldots,p\}$ the a_{i_k,j_k} lies in the right shadow of $a_{\alpha+1,\beta+1} = 0$ if $\alpha < \beta$, or in its left shadow if $\alpha > \beta$. This implies that

$$a_{i_k,j_k} = 0. \qquad \square$$

Totally positive matrices that satisfy the above property, i.e.,

$$A\begin{pmatrix} i_1,\ldots,i_p \\ j_1,\ldots,j_p \end{pmatrix} > 0$$

if and only if

$$a_{i_k,j_k} > 0, \qquad k = 1, \ldots, p,$$

are called *almost strictly totally positive* matrices.

1.4 Determinantal inequalities

There are various determinantal inequalities that hold for totally positive and strictly totally positive matrices. In this section we detail a few of them.

Generalized Hadamard Inequalities We start with the generalized Hadamard inequalities that hold for a wide class of matrices. In this monograph we prove these inequalities only for totally positive matrices. Let $\mathbf{i}$, $\mathbf{j}$, and $\mathbf{k}$ be disjoint sets in $\{1, \ldots, n\}$. We prove the following, where we slightly abuse notation.

Theorem 1.21 *Let A be any $n \times n$ totally positive matrix and $\mathbf{i}$, $\mathbf{j}$, and $\mathbf{k}$ be disjoint subsets of $\{1, \ldots, n\}$. Then,*

$$A(\mathbf{i}, \mathbf{j}, \mathbf{k})A(\mathbf{i}) \le A(\mathbf{i}, \mathbf{j})A(\mathbf{i}, \mathbf{k}). \tag{1.5}$$

If A is strictly totally positive, then strict inequality holds in (1.5).

We recall what we mean by the above inequality. If $\mathbf{i} = (i_1, \ldots, i_p) \in I^n_p$, $\mathbf{j} = (j_1, \ldots, j_q) \in I^n_q$, $\mathbf{k} = (k_1, \ldots, k_r) \in I^n_r$ are disjoint sets of ordered indices in $\{1, \ldots, n\}$, then the following inequalities hold between the principal minors of A:

$$A\begin{pmatrix} i_1, \ldots, i_p, j_1, \ldots, j_q, k_1, \ldots, k_r \\ i_1, \ldots, i_p, j_1, \ldots, j_q, k_1, \ldots, k_r \end{pmatrix} A\begin{pmatrix} i_1, \ldots, i_p \\ i_1, \ldots, i_p \end{pmatrix}$$
$$\le A\begin{pmatrix} i_1, \ldots, i_p, j_1, \ldots, j_q \\ i_1, \ldots, i_p, j_1, \ldots, j_q \end{pmatrix} A\begin{pmatrix} i_1, \ldots, i_p, k_1, \ldots, k_r \\ i_1, \ldots, i_p, k_1, \ldots, k_r \end{pmatrix},$$

where we (always) assume that the row and column indices have been rearranged in natural (increasing) order. This inequality is called a Generalized Hadamard Inequality. It is sometimes written

$$A(\mathbf{s} \cup \mathbf{t})A(\mathbf{s} \cap \mathbf{t}) \le A(\mathbf{s})A(\mathbf{t})$$

where $\mathbf{s}$ and $\mathbf{t}$ are arbitrary subsets of $\{1, \ldots, n\}$. In our notation $\mathbf{s} = \mathbf{i} \cup \mathbf{j}$ and $\mathbf{t} = \mathbf{i} \cup \mathbf{k}$. If $\mathbf{s} \cap \mathbf{t} = \mathbf{i} = \emptyset$, then (1.5) remains valid and is generally called a Hadamard Inequality. (For convenience we always set $A(\emptyset) = 1$.)

Proof We prove the result for a totally positive matrix. If $A(\mathbf{i},\mathbf{j},\mathbf{k}) = 0$ then (1.5) is certainly valid. As such we may assume that $A(\mathbf{i},\mathbf{j},\mathbf{k}) > 0$. Thus from Theorem 1.13 all the other minors in (1.5) are strictly positive.

Let

$$b_{k\ell} = A\begin{pmatrix}\mathbf{i}, k\\ \mathbf{i}, \ell\end{pmatrix}.$$

Set $|\mathbf{j}| = q$ and $|\mathbf{k}| = r$, where $|\cdot|$ denotes the cardinality of the set. From Sylvester's Determinant Identity

$$\begin{aligned} B(\mathbf{j}) &= A(\mathbf{i},\mathbf{j})A(\mathbf{i})^{q-1} \\ B(\mathbf{k}) &= A(\mathbf{i},\mathbf{k})A(\mathbf{i})^{r-1} \end{aligned}$$

and

$$B(\mathbf{j},\mathbf{k}) = A(\mathbf{i},\mathbf{j},\mathbf{k})A(\mathbf{i})^{q+r-1}.$$

Thus (1.5) is equivalent to the simpler Hadamard inequality

$$B(\mathbf{j},\mathbf{k}) \le B(\mathbf{j})B(\mathbf{k}), \tag{1.6}$$

where $\mathbf{j}$ and $\mathbf{k}$ are disjoint sets of ordered indices in $\{1,\ldots,n\}$, B is a totally positive matrix, and $B(\mathbf{j},\mathbf{k}) > 0$. It is this inequality that we now prove.

Our proof will be by induction on $q + r = m$. We always assume that $q, r \ge 1$. For $m = 2$ we must show that

$$B\begin{pmatrix}j, k\\ j, k\end{pmatrix} \le B\begin{pmatrix}j\\ j\end{pmatrix} B\begin{pmatrix}k\\ k\end{pmatrix} = b_{jj}b_{kk}$$

where $\mathbf{j} = (j)$ and $\mathbf{k} = (k)$. As

$$B\begin{pmatrix}j, k\\ j, k\end{pmatrix} = b_{jj}b_{kk} - b_{jk}b_{kj}$$

and $b_{jk}, b_{kj} \ge 0$, this inequality is immediate.

Assume $m > 2$ and $q > 1$. Let $j_1 \in \mathbf{j}$ and $\mathbf{j}' = \mathbf{j}\backslash\{j_1\}$ and set

$$c_{k\ell} = B\begin{pmatrix}j_1, k\\ j_1, \ell\end{pmatrix}$$

for $k, \ell \in \mathbf{j}' \cup \mathbf{k}$. From Sylvester's Determinant Identity, $C = (c_{k\ell})$ is totally positive and

$$C(\mathbf{j}', \mathbf{k}) > 0.$$

Now

$$\begin{aligned} C(\mathbf{j}') &= B(\mathbf{j})b_{j_1j_1}^{q-2} \\ C(\mathbf{k}) &= B(\mathbf{k}, j_1)b_{j_1j_1}^{r-1} \end{aligned}$$

and

$$C(\mathbf{j}', \mathbf{k}) = B(\mathbf{j}, \mathbf{k}) b_{j_1 j_1}^{q+r-2}.$$

From the induction hypothesis and since $|\mathbf{j}'| + |\mathbf{k}| = m - 1$

$$C(\mathbf{j}', \mathbf{k}) \leq C(\mathbf{j}')C(\mathbf{k}),$$

implying, by the above set of equalities,

$$B(\mathbf{j}, \mathbf{k}) b_{j_1 j_1} \leq B(\mathbf{j}) B(\mathbf{k}, j_1).$$

Since $|\mathbf{k}| + 1 = r + 1 < m$, we can again apply the induction hypothesis to obtain

$$B(\mathbf{k}, j_1) \leq B(\mathbf{k}) b_{j_1 j_1}.$$

Substituting in the above gives us the desired inequality (1.6).

This same proof, with minor modifications in order to verify strict inequality, is valid for strictly totally positive matrices. □

Remark In all the determinantal inequalities in this section we consider only principal minors. Recall (Proposition 1.2) that as all submatrices of (strictly) totally positive matrices are themselves (strictly) totally positive matrices these inequalities are also valid for nonprincipal minors. But be careful with the bookkeeping.

There are various consequences of this generalized Hadamard inequalities. We present two of them here.

Theorem 1.22 *Let A be an $n \times n$ totally positive matrix. Let $\mathbf{i}^1, \ldots, \mathbf{i}^q$ be arbitrary subsets of $\{1, \ldots, n\}$. Set*

$$\mathbf{j}^\ell = \{r : r \text{ belongs to at least } \ell \text{ of the } \mathbf{i}^1, \ldots, \mathbf{i}^q\}.$$

Then

$$A(\mathbf{j}^1) \cdots A(\mathbf{j}^q) \leq A(\mathbf{i}^1) \cdots A(\mathbf{i}^q).$$

If A is strictly totally positive, then strict inequality holds in the above.

Proof We prove this result for a totally positive matrix. The same proof, with minor modifications, is valid for strictly totally positive matrices. Our proof will be by induction on q. Note that the case $q = 2$ is exactly Theorem 1.21. Set

$$\mathbf{k}^\ell = \{r : r \text{ belongs to at least } \ell \text{ of the } \mathbf{i}^1, \ldots, \mathbf{i}^{q-1}\}.$$

From the induction hypothesis

$$A(\mathbf{k}^1)\cdots A(\mathbf{k}^{q-1}) \le A(\mathbf{i}^1)\cdots A(\mathbf{i}^{q-1}). \tag{1.7}$$

Let

$$\mathbf{m}^\ell = \{r : r \text{ belongs to } \mathbf{i}^q \text{ and at least } \ell \text{ of the } \mathbf{i}^1, \ldots, \mathbf{i}^{q-1}\}.$$

Thus

$$\mathbf{j}^\ell = \mathbf{k}^\ell \cup \mathbf{m}^{\ell-1} \text{ and } \mathbf{m}^\ell = \mathbf{k}^\ell \cap \mathbf{m}^{\ell-1}, \qquad \ell = 1, \ldots, q,$$

where $\mathbf{m}^0 = \mathbf{i}^q$, $\mathbf{m}^q = \mathbf{k}^q = \emptyset$ and $A(\emptyset) = 1$. Applying Theorem 1.21 we obtain

$$A(\mathbf{j}^\ell)A(\mathbf{m}^\ell) \le A(\mathbf{k}^\ell)A(\mathbf{m}^{\ell-1}), \qquad \ell = 1, \ldots, q.$$

Thus

$$\prod_{\ell=1}^{q} A(\mathbf{j}^\ell)A(\mathbf{m}^\ell) \le \prod_{\ell=1}^{q} A(\mathbf{k}^\ell)A(\mathbf{m}^{\ell-1}),$$

which reduces to

$$\prod_{\ell=1}^{q} A(\mathbf{j}^\ell) \le \left(\prod_{\ell=1}^{q-1} A(\mathbf{k}^\ell)\right) A(\mathbf{m}^0)$$

and from (1.7) this implies that

$$\prod_{\ell=1}^{q} A(\mathbf{j}^\ell) \le \left(\prod_{\ell=1}^{q-1} A(\mathbf{i}^\ell)\right) A(\mathbf{i}^q) = \prod_{\ell=1}^{q} A(\mathbf{i}^\ell).$$ □

Let A be an $n \times n$ matrix and for $r = 1, \ldots, n$, set

$$Q_r = \left[\prod_{1 \le i_1 < \cdots < i_r \le n} A\begin{pmatrix} i_1, \ldots, i_r \\ i_1, \ldots, i_r \end{pmatrix}\right]^{\frac{1}{\binom{n-1}{r-1}}}.$$

Thus

$$\begin{aligned} Q_1 &= \prod_{i=1}^{n} a_{ii} \\ Q_2 &= \left[\prod_{1 \le i_1 < i_2 \le n} A\begin{pmatrix} i_1, i_2 \\ i_1, i_2 \end{pmatrix}\right]^{\frac{1}{n-1}} \\ &\vdots \\ Q_n &= A\begin{pmatrix} 1, \ldots, n \\ 1, \ldots, n \end{pmatrix}. \end{aligned}$$

We prove that the Q_r are a decreasing sequence of values.

Theorem 1.23 *If A is an $n \times n$ totally positive matrix, then*

$$Q_n \le Q_{n-1} \le \cdots \le Q_1.$$

If A is strictly totally positive, then strict inequalities hold in the above.

Proof We prove this result for totally positive matrices. The same proof, with minor modifications, is valid for strictly totally positive matrices. Let

$$P_r = \left[\prod_{1 \le i_1 < \cdots < i_r \le n} A \begin{pmatrix} i_1, \ldots, i_r \\ i_1, \ldots, i_r \end{pmatrix} \right]^{\frac{1}{\binom{n}{r}}}, \qquad r = 1, \ldots, n,$$

and set $P_0 = 1$. It is more convenient to work with the P_r.

We prove that $P_{r+1}P_{r-1} \le P_r^2$ for $r = 1, \ldots, n-1$. Let us first consider the case $r = 1$, i.e., $P_2 \le P_1^2$. From the Hadamard inequality

$$A \begin{pmatrix} r, s \\ r, s \end{pmatrix} \le a_{rr} a_{ss}.$$

Thus, as is easily verified,

$$\prod_{1 \le r < s \le n} A \begin{pmatrix} r, s \\ r, s \end{pmatrix} \le \prod_{i=1}^{n} a_{ii}^{n-1}$$

from which

$$P_2 = \left[\prod_{1 \le r < s \le n} A \begin{pmatrix} r, s \\ r, s \end{pmatrix} \right]^{\frac{2}{n(n-1)}} \le \left[\prod_{i=1}^{n} a_{ii} \right]^{\frac{2}{n}} = P_1^2.$$

For $r = 2, \ldots, n-1$ the proof is similar, but the bookkeeping is more complicated. From the generalized Hadamard inequality we have

$$\begin{aligned} &A \begin{pmatrix} i_1, \ldots, i_{r-1}, i_r, i_{r+1} \\ i_1, \ldots, i_{r-1}, i_r, i_{r+1} \end{pmatrix} A \begin{pmatrix} i_1, \ldots, i_{r-1} \\ i_1, \ldots, i_{r-1} \end{pmatrix} \\ &\quad \le A \begin{pmatrix} i_1, \ldots, i_{r-1}, i_r \\ i_1, \ldots, i_{r-1}, i_r \end{pmatrix} A \begin{pmatrix} i_1, \ldots, i_{r-1}, i_{r+1} \\ i_1, \ldots, i_{r-1}, i_{r+1} \end{pmatrix}. \end{aligned}$$

Fixing $i_1, \ldots, i_{r-1}$ and taking products over $i_r < i_{r+1}$ (distinct from

$i_1, \ldots, i_{r-1}$) we obtain

$$\left[\prod_{i_r < i_{r+1}} A\begin{pmatrix} i_1, \ldots, i_{r-1}, i_r, i_{r+1} \\ i_1, \ldots, i_{r-1}, i_r, i_{r+1} \end{pmatrix}\right] \left[A\begin{pmatrix} i_1, \ldots, i_{r-1} \\ i_1, \ldots, i_{r-1} \end{pmatrix}\right]^{\binom{n-r+1}{2}}$$
$$\leq \left[\prod_{i_r} A\begin{pmatrix} i_1, \ldots, i_{r-1}, i_r \\ i_1, \ldots, i_{r-1}, i_r \end{pmatrix}\right]^{n-r}.$$

We now vary over $i_1, \ldots, i_{r-1}$ to get

$$\left[\prod_{1 \leq i_1 < \cdots < i_{r+1} \leq n} A\begin{pmatrix} i_1, \ldots, i_{r-1}, i_r, i_{r+1} \\ i_1, \ldots, i_{r-1}, i_r, i_{r+1} \end{pmatrix}\right]^{\binom{r+1}{r-1}} \times$$
$$\left[\prod_{1 \leq i_1 < \cdots < i_{r-1} \leq n} A\begin{pmatrix} i_1, \ldots, i_{r-1} \\ i_1, \ldots, i_{r-1} \end{pmatrix}\right]^{\binom{n-r+1}{2}}$$
$$\leq \left[\prod_{1 \leq i_1 < \cdots < i_r \leq n} A\begin{pmatrix} i_1, \ldots, i_r \\ i_1, \ldots, i_r \end{pmatrix}\right]^{(n-r)\binom{r}{r-1}},$$

i.e.,

$$P_{r+1}^{\binom{n}{r+1}\binom{r+1}{r-1}} P_{r-1}^{\binom{n}{r-1}\binom{n-r+1}{2}} \leq P_r^{\binom{n}{r}(n-r)\binom{r}{r-1}}.$$

A simple calculation reduces this to

$$P_{r+1}P_{r-1} \leq P_r^2.$$

Now

$$Q_r^{r/n} = P_r, \qquad r = 1, \ldots, n.$$

Thus we have

$$Q_{r+1}^{r+1}Q_{r-1}^{r-1} \leq Q_r^{2r}, \qquad r = 1, \ldots, n-1,$$

where $Q_0 = 1$. For $r = 1$ this is

$$Q_2^2 \leq Q_1^2$$

implying

$$Q_2 \leq Q_1.$$

Assume $2 \leq r \leq n-1$ and

$$Q_r \leq \cdots \leq Q_1.$$

From the above

$$Q_{r+1}^{r+1}Q_{r-1}^{r-1} \le Q_r^{2r} \le Q_r^{r+1}Q_{r-1}^{r-1}$$

and therefore

$$Q_{r+1} \le Q_r.$$

Thus

$$Q_n \le \cdots \le Q_1. \qquad \square$$

There are other determinantal inequalities valid for totally positive matrices. We here note one such additional inequality.

Proposition 1.24 *Assume $A = (a_{ij})$ is an $n \times n$ strictly totally positive matrix, and we are given $0 \le j_i \le n - i$, $i = 1, \ldots, n-1$, with*

$$0 \le j_{n-1} \le \cdots \le j_1.$$

Define the $n \times n$ matrix $C = (c_{ij})$ where, for each i,

$$c_{ij} = \begin{cases} 0, & j = 1, \ldots, j_i \\ a_{ij}, & j = j_i + 1, \ldots, n. \end{cases}$$

Then

$$(-1)^{j_1 + \cdots + j_{n-1}} \det C > 0.$$

Proof We prove this result by induction and by the use of Sylvester's Determinant Identity. It is easily seen to hold for $n = 2$.

If $j_k = n - k$ for some k, then

$$\det C = (-1)^{k(n-k)} C\begin{pmatrix} 1, \ldots, k \\ n-k+1, \ldots, n \end{pmatrix} C\begin{pmatrix} k+1, \ldots, n \\ 1, \ldots, n-k \end{pmatrix}$$

and one can apply induction directly to obtain our result.

Let us therefore assume that $n \ge 3$, the j_i are nonincreasing in i, and $0 \le j_i < n - i$, $i = 1, \ldots, n-2$ (note that $j_{n-1} = 0$). Set

$$b_{ij} = C\begin{pmatrix} 2, \ldots, n-1, i \\ 2, \ldots, n-1, j \end{pmatrix}, \qquad i, j \in \{1, n\}.$$

Thus, from Sylvester's Determinant Identity,

$$b_{11}b_{nn} - b_{1n}b_{n1} = C\begin{pmatrix} 2, \ldots, n-1 \\ 2, \ldots, n-1 \end{pmatrix} C\begin{pmatrix} 1, \ldots, n \\ 1, \ldots, n \end{pmatrix}.$$

Let r denote the number of zeros in the first column of C. Then by the induction hypothesis

$$(-1)^{j_2+\cdots+j_{n-2}-(r-1)}C\begin{pmatrix}2,\ldots,n-1\\2,\ldots,n-1\end{pmatrix}>0,$$

and applying the induction hypothesis to each of the b_{ij} we get,

$$\begin{aligned}(-1)^{j_1+\cdots+j_{n-2}}b_{11} &= (-1)^{j_1+\cdots+j_{n-2}}C\begin{pmatrix}1,\ldots,n-1\\1,\ldots,n-1\end{pmatrix}>0,\\(-1)^{j_2+\cdots+j_{n-2}-(r-1)}b_{nn} &= (-1)^{j_2+\cdots+j_{n-2}-(r-1)}C\begin{pmatrix}2,\ldots,n\\2,\ldots,n\end{pmatrix}>0,\\(-1)^{j_1+\cdots+j_{n-2}-r}b_{1n} &= (-1)^{j_1+\cdots+j_{n-2}-r}C\begin{pmatrix}1,\ldots,n-1\\2,\ldots,n\end{pmatrix}>0,\\(-1)^{j_2+\cdots+j_{n-2}}b_{n1} &= (-1)^{j_2+\cdots+j_{n-2}}C\begin{pmatrix}2,\ldots,n\\1,\ldots,n-1\end{pmatrix}>0.\end{aligned}$$

Thus

$$(-1)^{j_1-(r-1)}[b_{11}b_{nn}-b_{1n}b_{n1}]>0$$

and therefore

$$(-1)^{j_1+\cdots+j_{n-1}}\det C>0.$$ □

If A is only totally positive then the strict inequality in the statement of Proposition 1.24 should be replaced by an inequality. This is most easily verified by an appeal to Theorem 2.6 of the next chapter.

We now apply Proposition 1.24 (after interchanging row and column indices) in order to prove Theorem 1.8.

Proof of Theorem 1.8 Recall that

$$b_{ij}=a_{ij}A\begin{pmatrix}1,\ldots,k\\1,\ldots,k\end{pmatrix}-A\begin{pmatrix}1,\ldots,k,i\\1,\ldots,k,j\end{pmatrix}$$

for $i=k+1,\ldots,n$, $j=k+1,\ldots,m$. Expand the rightmost minor by its last row to obtain

$$b_{ij}=\sum_{s=1}^{k}a_{is}(-1)^{s+k}A\begin{pmatrix}1,\ldots,k\\1,\ldots,\widehat{s},\ldots,k,j\end{pmatrix}.$$

From this expression we see that rank $B\le k$.

We shall prove that if A is strictly totally positive then all $r\times r$ minors of B are strictly positive for $r=1,\ldots,\min\{k,n-k,m-k\}$. (By appealing to Theorem 2.6 we then can show that if A is only totally positive then B is totally positive.)

Set

$$c_{sj} = (-1)^{s+k} A \begin{pmatrix} 1,\ldots,k \\ 1,\ldots,\widehat{s},\ldots,k,j \end{pmatrix}, \qquad s = 1,\ldots,k,\ j = k+1,\ldots,m.$$

Thus for $r \leq \min\{k, n-k, m-k\}$ we have

$$B\begin{pmatrix} i_1,\ldots,i_r \\ j_1,\ldots,j_r \end{pmatrix} = \sum_{1\leq s_1<\cdots<s_r\leq k} A\begin{pmatrix} i_1,\ldots,i_r \\ s_1,\ldots,s_r \end{pmatrix} C\begin{pmatrix} s_1,\ldots,s_r \\ j_1,\ldots,j_r \end{pmatrix}$$

by the Cauchy–Binet formula.

From Theorem 1.12

$$\begin{aligned} C\begin{pmatrix} s_1,\ldots,s_r \\ j_1,\ldots,j_r \end{pmatrix} &= (-1)^{kr+\frac{r(r-1)}{2}+\sum_{\ell=1}^{r} s_\ell} \left[A\begin{pmatrix} 1,\ldots,k \\ 1,\ldots,k \end{pmatrix}\right]^{r-1} \\ &\qquad A\begin{pmatrix} 1,\ldots,k \\ s_1',\ldots,s_{k-r}' j_1,\ldots,j_r \end{pmatrix}, \end{aligned}$$

where $s_1',\ldots,s_{k-r}'$ is complementary to $s_1,\ldots,s_r$ in $\{1,\ldots,k\}$.

Therefore

$$\begin{aligned} & B\begin{pmatrix} i_1,\ldots,i_r \\ j_1,\ldots,j_r \end{pmatrix} \\ &= \sum_{1\leq s_1<\cdots<s_r\leq k} A\begin{pmatrix} i_1,\ldots,i_r \\ s_1,\ldots,s_r \end{pmatrix} C\begin{pmatrix} s_1,\ldots,s_r \\ j_1,\ldots,j_r \end{pmatrix} \\ &= (-1)^{kr+\frac{r(r-1)}{2}} \left[A\begin{pmatrix} 1,\ldots,k \\ 1,\ldots,k \end{pmatrix}\right]^{r-1} \times \\ &\quad \sum_{1\leq s_1<\cdots<s_r\leq k} (-1)^{\sum_{\ell=1}^{r} s_\ell} A\begin{pmatrix} i_1,\ldots,i_r \\ s_1,\ldots,s_r \end{pmatrix} A\begin{pmatrix} 1,\ldots,k \\ s_1',\ldots,s_{k-r}',j_1,\ldots,j_r \end{pmatrix}. \end{aligned}$$

We now apply Proposition 1.24 to the $(k+r)\times(k+r)$ strictly totally positive matrix

$$A\begin{bmatrix} 1,\ldots,k,i_1,\ldots,i_r \\ 1,\ldots,k,j_1,\ldots,j_r \end{bmatrix}$$

where we set as zero the $r\times r$ bottom right corner of this matrix, i.e., set as zero the (i,j) entries for $i,j = k+1,\ldots,k+r$. It follows from Proposition 1.3 that Proposition 1.24 also holds in this case as interchanging all rows and columns preserves strict total positivity. The resulting matrix has sign

$(-1)^r$. Now apply Laplace's expansion by minors to this matrix to obtain

$$(-1)^r \sum_{1 \le s_1 < \cdots < s_r \le k} (-1)^{(k+1+s_1)+\cdots+(k+r+s_r)} A\begin{pmatrix} i_1, \ldots, i_r \\ s_1, \ldots, s_r \end{pmatrix} \\ \times A\begin{pmatrix} 1, \ldots, k \\ s'_1, \ldots, s'_{k-r}, j_1, \ldots, j_r \end{pmatrix} > 0.$$

Thus

$$B\begin{pmatrix} i_1, \ldots, i_r \\ j_1, \ldots, j_r \end{pmatrix} > 0. \qquad \square$$

1.5 Remarks

The study of total positivity and strict total positivity for continuous kernels predates the study of total positivity and strict total positivity for matrices (see e.g., Kellogg [1918]). Such phenomena are not uncommon. The three pioneers of the theory of totally positive matrices are F. R. Gantmacher, M. G. Krein, and I. J. Schoenberg. It was Schoenberg [1930] who first considered such matrices. He did so in his study of variation diminishing properties (see Chapter 3). In fact Schoenberg also coined the term *total positiv* (in German) in his 1930 paper. Krein came to consider such kernels and (later) matrices as a consequence of his research on ordinary differential equations whose Green's kernel is totally positive. The joint paper of Gantmacher, Krein [1937] (an announcement appeared in Gantmacher, Krein [1935]) presented most of the main results relating to spectral properties of totally positive matrices, and many other important results concerning totally positive matrices (except that they were then unaware of Schoenberg's earlier paper and the variation diminishing properties associated with such matrices). This 1937 paper, in a slightly expanded form, is most of Chapter II of the book Gantmacher, Krein [1950]. In Gantmacher, Krein [1937] they used the term *complètement non négative* and *complètement positive* (French) for totally positive and strictly totally positive, respectively. As such many authors use the terms *totally nonnegative* and *totally positive* for totally positive and strictly totally positive, respectively.

For a proof and history of the Cauchy–Binet formula, Sylvester's Determinant Identity and the Laplace expansion by minors, see Brualdi, Schneider [1983] and references therein. The initial propositions of Section 1.2 can, for the most part, be found in Gantmacher, Krein [1937], Ando [1987], and Karlin [1968]. Theorem 1.8 is from Ando [1987], Theorem 3.9. The proof as presented here is very different. Proposition 1.9

is due to Whitney [1952]. Theorem 1.11 is in Karlin [1968], Theorem 6.1, p. 132. The construction at the end of Section 1.2 is from Boocher, Froehle [2008]. Theorem 1.13 is contained in Karlin [1968], p. 89. The proof therein is different. Proposition 1.15 is from de Boor, Pinkus [1982] (see also Koteljanskii [1950]). Propositions 1.16 and 1.17 are from Gantmacher, Krein [1937], and can also be found in Karlin [1968]. Theorem 1.19 is in Pinkus [2008].

Almost strictly totally positive matrices were introduced in Gasca, Micchelli, Peña [1992]. Proposition 1.20 can be found there, although the proof is much different.

The simplest form of the Hadamard Inequality for totally positive matrices can already be found in Gantmacher, Krein [1937], p. 450. The form of the generalized Hadamard inequalities as given here is from Koteljanskii [1950], where general conditions are given for a generalized Hadamard inequality to hold. Generalized Hadamard inequalities hold for positive definite matrices and M-matrices. A square matrix is an M-matrix if all its principal minors are positive and all its off-diagonal entries are nonpositive.

Theorem 1.22 is contained in Carlson [1968]. A generalization to

$$\prod_{k=r}^{q} [A(\mathbf{j}^k)]^{\binom{k-1}{r-1}} \le \prod_{1\le \ell_1<\cdots<\ell_r\le q} A(\mathbf{i}^{\ell_1}\cap\cdots\cap\mathbf{i}^{\ell_r})$$

for any $r \le q$ may be found in Fan [1968]. (The case $r = 1$ is Theorem 1.21.) Theorem 1.23 for positive definite matrices is known as Szasz's inequality (see Beckenbach, Bellman [1961], p. 64). It only depends upon the nonnegativity of the principal minors and the generalized Hadamard inequalities and thus is also valid for totally positive matrices, as was noted by Koteljanskii [1950]. It was generalized, for this same class of matrices, in Fan [1967] to the following. Let A be an $n\times n$ matrix and let $\mathbf{s}^1,\ldots,\mathbf{s}^q$ be pairwise disjoint subsets of $\{1,\ldots,n\}$, $q\ge 2$. Set

$$Q_r = \left[\prod_{1\le i_1<\cdots<i_r\le q} A\left(\mathbf{s}^{i_1}\cup\cdots\cup\mathbf{s}^{i_r}\right)\right]^{\frac{1}{\binom{q-1}{r-1}}}.$$

Then

$$Q_q \le \cdots \le Q_1.$$

Other determinantal inequalities exist for (strictly) totally positive matrices (see e.g., Mühlbach, Gasca [1985], Fallat, Gekhtman, Johnson [2003], Skandera [2004] and Boocher, Froehle [2008]). The construction at the end of Section 1.2 permits us to generalize many of the determinantal inequalities of Section 1.4 (there must exist the same number of factors on both sides of the inequality) to where we have nonprincipal minors (see Boocher, Froehle [2008]).

Boocher, Froehle [2008], building on work of Skandera [2004] and Fallat, Gekhtman, Johnson [2003], were able to characterize all determinantal inequalities of the form

$$A\begin{pmatrix} i_1,\ldots,i_k \\ j_1,\ldots,j_k \end{pmatrix} A\begin{pmatrix} m_1,\ldots,m_p \\ \ell_1,\ldots,\ell_p \end{pmatrix} \le A\begin{pmatrix} r_1,\ldots,r_t \\ s_1,\ldots,s_t \end{pmatrix} A\begin{pmatrix} u_1,\ldots,u_q \\ v_1,\ldots,v_q \end{pmatrix} \tag{1.8}$$

valid for all $n \times n$ totally positive matrices. The analogous problem for more than two factors has yet to be solved, although numerous results can be found in these papers.

One characterization of the inequalities (1.8) is the following that appears in Boocher, Froehle [2008]. For ease of exposition we introduce the following notation. For each $\alpha = (\alpha_1,\ldots,\alpha_n)$, a vector of n distinct integers in $\{1,\ldots,2n\}$, we let

$$A(\alpha) = A\begin{pmatrix} i_1,\ldots,i_k \\ j_1,\ldots,j_k \end{pmatrix},$$

where $\{\alpha_1,\ldots,\alpha_n\} = \{i_1,\ldots,i_k, 2n+1-j'_1,\ldots,2n+1-j'_{n-k}\}$, and $\{j'_1,\ldots,j'_{n-k}\}$ is complementary to $\{j_1,\ldots,j_k\}$ in $\{1,\ldots,n\}$. For $k=0$, i.e., $\alpha = (n+1,\ldots,2n)$, we set $A(\alpha)=1$. Based on results of this chapter we have shown or can easily show that

$$A(i,j+1,\Delta)A(i+1,j,\Delta) \le A(i,j,\Delta)A(i+1,j+1,\Delta) \tag{1.9}$$

where $i,j \in \{1,\ldots,2n\}$, $\Delta \subset \{1,\ldots,2n\}$ with $|\Delta| = n-2$, $i+1, j+1$ are understood mod $2n$, and $i, i+1, j, j+1, \Delta$ are all distinct. Boocher, Froehle [2008] proved that a determinantal inequality of the form (1.8) holds for all $n \times n$ totally positive matrices if and only if the inequality can be factored as a product of inequalities of the form (1.9), where we vary over i, j, and Δ as above. Other characterizations may be found in Boocher, Froehle [2008].

2
Criteria for total positivity and strict total positivity

In this chapter we discuss the problem of establishing determinantal criteria for when a matrix is totally positive or strictly totally positive. Totally different criteria will be discussed in the next chapter on variation diminishing.

In Section 2.1 we prove Fekete's Lemma and some of its consequences. The most notable thereof is Theorem 2.3, which states that for a matrix to be strictly totally positive it suffices to prove that all its minors composed of the first k rows and k consecutive columns and all its minors composed of the first k columns and k consecutive rows are strictly positive for all possible k. We apply the results of Section 2.1 in Section 2.2, where we prove that strictly totally positive matrices are dense in the class of totally positive matrices, and provide proofs of Propositions 1.9 and 1.10 from Chapter 1.

In Section 2.3 we discuss triangular matrices, detail determinantal criteria for when such matrices are totally positive, and in Section 2.4 we consider the LDU-factorization of strictly totally positive and totally positive matrices. We will return to the study of factorizations of totally positive matrices in Chapter 6. In Section 2.5 we consider determinantal criteria for when a matrix is totally positive. The results are nowhere near as elegant as those valid for strictly totally positive matrices.

In Section 2.6 we prove a recent surprising and beautiful result of O. M. Katkova and A. M. Vishnyakova (completing work initiated by T. Craven and G. Csordas). They prove that if $A = (a_{ij})$ is an $n \times n$ matrix, all of whose entries are strictly positive, and if

$$a_{ij}a_{i+1,j+1} > \left(4\cos^2 \frac{\pi}{n+1}\right) a_{i,j+1}a_{i+1,j}$$

for all $i, j = 1, \ldots, n-1$, then A is strictly totally positive. Furthermore, the constant $4\cos^2 \frac{\pi}{n+1}$ in the above inequality is best possible.

2.1 Criteria for strict total positivity

A matrix $A = (a_{ij})_{i=1}^{n}{}_{j=1}^{m}$ is strictly totally positive if and only if

$$A\begin{pmatrix} i_1, \ldots, i_p \\ j_1, \ldots, j_p \end{pmatrix} > 0$$

for all $1 \le i_1 < \cdots < i_p \le n$, $1 \le j_1 < \cdots < j_p \le m$, and all $p = 1, \ldots, \min\{n, m\}$. Thus, formally, we must verify

$$\sum_{p=1}^{\min\{n,m\}} \binom{n}{p}\binom{m}{p} = \binom{n+m}{n} - 1$$

conditions. For $n = m$ this number is on the order of $4^n n^{-1/2}$. Thankfully, it is not necessary to check all these conditions. For example, if $a_{i1} > 0$ for $i = 1, \ldots, n$, $a_{1j} > 0$ for $j = 1, \ldots, m$, and

$$A\begin{pmatrix} 1, i \\ 1, j \end{pmatrix} = a_{11}a_{ij} - a_{i1}a_{1j} > 0,$$

for $i = 2, \ldots, n$ and $j = 2, \ldots, m$, then we immediately obtain $a_{ij} > 0$ for all $i, j > 1$. In other words, many of the inequalities in the definition of a strictly totally positive matrix are superfluous. In this section we search for reasonable determinantal criteria establishing strict total positivity.

We start with a classic result due to Fekete.

Lemma 2.1 (Fekete's Lemma). *Assume C is an $n \times m$ matrix ($n \ge m$) such that all $(m-1)$st order minors with columns $1, \ldots, m-1$ are strictly positive and the mth order minors composed from consecutive rows are also strictly positive. Then all mth order minors of C are strictly positive.*

Proof For a given strictly increasing sequence $\mathbf{i} = (i_1, \ldots, i_k)$ of k integers in $\{1, \ldots, n\}$, i.e., $\mathbf{i} \in I_k^n$, we define $d(\mathbf{i})$, the *dispersion* of $\mathbf{i}$, by

$$d(\mathbf{i}) := \sum_{j=2}^{k} (i_j - i_{j-1} - 1) = (i_k - i_1) - (k-1) \ge 0.$$

$d(\mathbf{i})$ counts the number of integers between i_1 and i_k that are not in the sequence $i_1, \ldots, i_k$. Thus $d(\mathbf{i}) = 0$ if and only if the sequence $\mathbf{i}$ is composed of consecutive integers. We will prove this lemma by induction on $d(\cdot)$.

Let $\mathbf{i} = (i_1, \ldots, i_m)$ be as above. We wish to prove that

$$C\begin{pmatrix} i_1, \ldots, i_m \\ 1, \ldots, m \end{pmatrix} > 0.$$

By our assumption, this is valid when $d(\mathbf{i}) = 0$.

For $n = m$ there is nothing to prove, and as such we assume that $n > m$. Let $\mathbf{i} = (i_1, \ldots, i_m)$ with $d(\mathbf{i}) = r > 0$, and assume the result holds for all $\mathbf{i}$ with $d(\mathbf{i}) < r$. As the sequence $\mathbf{i}$ is not composed of consecutive integers, we can add to this sequence an integer between i_1 and i_m. Let $1 \le j_1 < \cdots < j_{m+1} \le n$ be the resulting sequence assuming j_ℓ to be the new index.

Consider the determinantal identity (1.2) from Chapter 1:

$$\begin{aligned} C&\begin{pmatrix} j_2, \ldots, j_m \\ 1, \ldots, m-1 \end{pmatrix} C\begin{pmatrix} j_1, \ldots, \widehat{j_\ell}, \ldots, j_{m+1} \\ 1, \ldots, m \end{pmatrix} \\ &= C\begin{pmatrix} j_1, \ldots, \widehat{j_\ell}, \ldots, j_m \\ 1, \ldots, m-1 \end{pmatrix} C\begin{pmatrix} j_2, \ldots, j_{m+1} \\ 1, \ldots, m \end{pmatrix} \\ &\quad + C\begin{pmatrix} j_2, \ldots, \widehat{j_\ell}, \ldots, j_{m+1} \\ 1, \ldots, m-1 \end{pmatrix} C\begin{pmatrix} j_1, \ldots, j_m \\ 1, \ldots, m \end{pmatrix}. \end{aligned}$$

In the above equation the three $m-1$ order minors are all positive by assumption, as they are all based on the columns $1, \ldots, m-1$. Now $d(\mathbf{i}) = (i_m - i_1) - (m-1) = (j_{m+1} - j_1) - (m-1) = r$, while $(j_{m+1} - j_2) - (m-1)$ and $(j_m - j_1) - (m-1)$ are strictly less than r. Thus, by our induction hypothesis, we have

$$C\begin{pmatrix} j_2, \ldots, j_{m+1} \\ 1, \ldots, m \end{pmatrix}, \quad C\begin{pmatrix} j_1, \ldots, j_m \\ 1, \ldots, m \end{pmatrix} > 0.$$

This implies that

$$C\begin{pmatrix} j_1, \ldots, \widehat{j_\ell}, \ldots, j_{m+1} \\ 1, \ldots, m \end{pmatrix} = C\begin{pmatrix} i_1, \ldots, i_m \\ 1, \ldots, m \end{pmatrix} > 0. \qquad \square$$

We will use the method of proof of Fekete's Lemma at least twice more in this chapter. An immediate consequence of the above result is the following.

Theorem 2.2 *Let A be an $n \times m$ matrix. Assume that all kth order minors of A composed from consecutive rows and consecutive columns are strictly positive for $k = 1, \ldots, \min\{n, m\}$. Then A is strictly totally positive.*

However, we need not verify all these inequalities in order to determine if A is strictly totally positive. It suffices to check the strict positivity of an even smaller number of determinants.

Theorem 2.3 *Let A be an $n \times m$ matrix. Assume that all kth order minors of A composed from the first k rows and k consecutive columns, and also all kth order minors of A composed from the first k columns and k consecutive rows, are strictly positive for $k = 1, \ldots, \min\{n, m\}$. Then A is strictly totally positive.*

There are thus exactly nm determinantal inequalities to be checked. For $n = m$ this n^2 is a significantly smaller value than $4^n n^{-1/2}$.

Proof We prove that the conditions in Theorem 2.3 imply that all kth order minors of A composed from consecutive rows and columns are strictly positive for $k = 1, \ldots, \min\{n, m\}$. We then apply Theorem 2.2 to obtain our desired result. In other words we prove that

$$A\begin{pmatrix} i+1, \ldots, i+k \\ j+1, \ldots, j+k \end{pmatrix} > 0 \tag{2.1}$$

for $i = 0, \ldots, n-k$, $j = 0, \ldots, m-k$, and $k = 1, \ldots, \min\{n, m\}$. Our assumption is that this result holds when $i = 0$ and when $j = 0$.

Let us first prove that (2.1) also holds when $j = 1$, i.e.,

$$A\begin{pmatrix} i+1, \ldots, i+k \\ 2, \ldots, k+1 \end{pmatrix} > 0.$$

We prove this by induction on i and on k. Assume $i = 1$. For $k = 1$ we must prove that

$$A\begin{pmatrix} 2 \\ 2 \end{pmatrix} = a_{22} > 0.$$

Now

$$A\begin{pmatrix} 1,2 \\ 1,2 \end{pmatrix} = a_{11}a_{22} - a_{21}a_{12} > 0.$$

By assumption $a_{11}, a_{21}, a_{12} > 0$. Thus $a_{22} > 0$. Assume the result holds for $k-1$. By an application of Sylvester's Determinant Identity,

$$A\begin{pmatrix} 1, \ldots, k \\ 1, \ldots, k \end{pmatrix} A\begin{pmatrix} 2, \ldots, k+1 \\ 2, \ldots, k+1 \end{pmatrix} - A\begin{pmatrix} 1, \ldots, k \\ 2, \ldots, k+1 \end{pmatrix} A\begin{pmatrix} 2, \ldots, k+1 \\ 1, \ldots, k \end{pmatrix}$$
$$= A\begin{pmatrix} 2, \ldots, k \\ 2, \ldots, k \end{pmatrix} A\begin{pmatrix} 1, \ldots, k+1 \\ 1, \ldots, k+1 \end{pmatrix}.$$

All determinants containing initial consecutive rows or columns are strictly positive by assumption, and from the induction hypothesis

$$A\begin{pmatrix}2,\ldots,k\\2,\ldots,k\end{pmatrix} > 0.$$

Thus

$$A\begin{pmatrix}2,\ldots,k+1\\2,\ldots,k+1\end{pmatrix} > 0.$$

Let us now consider the case $i = 2$. For $k = 1$ we must prove that $a_{32} > 0$. Now

$$A\begin{pmatrix}2,3\\1,2\end{pmatrix} = a_{21}a_{32} - a_{31}a_{22} > 0.$$

By assumption $a_{21}, a_{31} > 0$. From the previous analysis $a_{22} > 0$. Thus $a_{32} > 0$. Assume the result holds for $k-1$. From Sylvester's Determinant Identity,

$$A\begin{pmatrix}2,\ldots,k+1\\1,\ldots,k\end{pmatrix} A\begin{pmatrix}3,\ldots,k+2\\2,\ldots,k+1\end{pmatrix} - A\begin{pmatrix}2,\ldots,k+1\\2,\ldots,k+1\end{pmatrix} A\begin{pmatrix}3,\ldots,k+2\\1,\ldots,k\end{pmatrix}$$
$$= A\begin{pmatrix}3,\ldots,k+1\\2,\ldots,k\end{pmatrix} A\begin{pmatrix}2,\ldots,k+2\\1,\ldots,k+1\end{pmatrix}.$$

All determinants containing initial consecutive rows or columns are strictly positive, and by the above case $i = 1$

$$A\begin{pmatrix}2,\ldots,k+1\\2,\ldots,k+1\end{pmatrix} > 0.$$

From the induction hypothesis

$$A\begin{pmatrix}3,\ldots,k+1\\2,\ldots,k\end{pmatrix} > 0.$$

Thus

$$A\begin{pmatrix}3,\ldots,k+2\\2,\ldots,k+1\end{pmatrix} > 0.$$

We continue this process for all possible i. We do the same for $i = 1$ and all j and k. We then go through the argument again for $j = 2$, then $i = 2$, etc..., or alternatively, since the above result now holds for $i = 1$ and $j = 1$, we can apply an induction argument to the $(n-1) \times (m-1)$ matrix obtained from A by deleting its first row and column. □

In Section 2.4 of this chapter we give a different and more transparent proof of this result.

When dealing with strictly totally positive matrices we can reverse the order of the rows and columns (see Proposition 1.3). As a consequence thereof, we have the following.

Corollary 2.4 *Let A be an $n \times m$ matrix. Assume that all kth order minors of A composed from the last k rows and k consecutive columns, and also all kth order minors of A composed from the last k columns and k consecutive rows, are strictly positive for $k = 1, \ldots, \min\{n, m\}$. Then A is strictly totally positive.*

Similar corollaries hold for many of the results of this chapter. They should be understood. We will not bother to state them again.

Is this the minimal number of determinants that must be checked when determining whether A is a strictly totally positive matrix? It seems that it must be minimal as the number of conditions is nm, which is the number of entries in the matrix A.

The following is presented here as it fits in with our present discussion and can be seen to be a simple consequence of Theorem 2.3. It is also an immediate corollary of Theorem 1.19. We state it without proof.

Proposition 2.5 *Let A be an $n \times n$ totally positive matrix. Then A is strictly totally positive if*

$$A\begin{pmatrix} 1, \ldots, k \\ n-k+1, \ldots, n \end{pmatrix} > 0$$

$$A\begin{pmatrix} n-k+1, \ldots, n \\ 1, \ldots, k \end{pmatrix} > 0$$

for $k = 1, \ldots, n$.

2.2 Density and some further applications

Strictly totally positive matrices are easier to handle than totally positive matrices and fortuitously the class of strictly totally positive matrices is dense in the class of all totally positive matrices. That is, if A is an $n \times m$ totally positive matrix, then there are sequences $(A_k)_{k=1}^{\infty}$ of $n \times m$ strictly totally positive matrices such that elementwise

$$\lim_{k \to \infty} A_k = A.$$

This result will prove very useful.

Theorem 2.6 *Strictly totally positive matrices are dense in the class of totally positive matrices.*

Proof For each $q \in (0,1)$ the matrix $Q_n(q) = (q^{(i-j)^2})_{i=1}^{n}{}_{j=1}^{n}$ is a strictly totally positive matrix. This may be proven as follows. As $q^{(i-j)^2} = q^{i^2}q^{j^2}q^{-2ij}$ it suffices to prove that p^{ij} is strictly totally positive for $p = q^{-2} > 1$. Let $P = (p^{ij})_{i=1}^{n}{}_{j=1}^{m}$. Consider

$$P\begin{pmatrix} i+1,\ldots,i+k \\ j+1,\ldots,j+k \end{pmatrix}.$$

Set $x_r = p^{i+r}$, $r = 1,\ldots,k$. It then easily follows, factoring out from the rth row of this minor $p^{(i+r)(j+1)}$, that

$$P\begin{pmatrix} i+1,\ldots,i+k \\ j+1,\ldots,j+k \end{pmatrix} = \left(\prod_{r=1}^{k} p^{(i+r)(j+1)}\right) V\left(x_1,\ldots,x_k\right),$$

where $V(x_1,\ldots,x_k)$ is the Vandermonde of the points $x_1,\ldots,x_k$ and thus equals

$$\prod_{1\le s<t\le k}(x_t - x_s) = \prod_{1\le s<t\le k}(p^{i+t} - p^{i+s}).$$

The terms in the product on the right are strictly positive since $p > 1$. This implies that

$$P\begin{pmatrix} i+1,\ldots,i+k \\ j+1,\ldots,j+k \end{pmatrix} > 0$$

and therefore, from Theorem 2.2, the matrix $Q_n(q)$ is strictly totally positive. The matrix $Q_n(q)$ has the further property that

$$\lim_{q\to 0^+} Q_n(q) = I_n$$

(elementwise) where I_n is the $n \times n$ identity matrix.

Now, let A be an arbitrary $n \times m$ totally positive matrix of rank r. Form

$$B_q = Q_n(q)AQ_m(q).$$

Then each B_q is of rank r, and as is easily verified using the Cauchy–Binet formula, all the $k \times k$ minors of B_q are strictly positive for every $k \le r$. Furthermore

$$\lim_{q\to 0^+} B_q = A.$$

If $r = \min\{n,m\}$ we are finished. Assume not and let B'_q be identical with the matrix B_q except that we add an arbitrarily small $\varepsilon > 0$ to the

(1, 1)-element of B'_q. The matrix B'_q is totally positive and of rank $r+1$. As such, by the same method as employed above, we can approximate B'_q arbitrarily well by a matrix of rank $r+1$, all of whose $k \times k$ minors are strictly positive for every $k \le r+1$. We continue this process to obtain the result. □

Determinantal criteria for when a matrix is totally positive will be discussed in Section 2.4. But we present here a particular case of Theorem 2.13 since it can be easily proven by an application of Theorem 2.2 and the method of proof of Theorem 2.6. This is not the best possible result, as will be shown in Proposition 2.15.

Proposition 2.7 *Let A be an $n \times n$ nonsingular matrix. Assume that all kth order minors of A composed from arbitrary k rows and k consecutive columns, or all kth order minors of A composed from arbitrary k columns and k consecutive rows are nonnegative for $k = 1, \ldots, n$. Then A is totally positive.*

Proof Assume all kth order minors of A composed from arbitrary k rows and k consecutive columns are nonnegative. Let $Q(q)$ be the $n \times n$ matrix as used in the proof of Theorem 2.6, $q \in (0, 1)$. Set

$$A_q = Q(q)A.$$

From the Cauchy–Binet formula

$$A_q \begin{pmatrix} i+1, \ldots, i+k \\ j+1, \ldots, j+k \end{pmatrix} = \sum_{1 \le \ell_1 < \cdots < \ell_k \le n} Q(q) \begin{pmatrix} i+1, \ldots, i+k \\ \ell_1, \ldots, \ell_k \end{pmatrix} A \begin{pmatrix} \ell_1, \ldots, \ell_k \\ j+1, \ldots, j+k \end{pmatrix}.$$

By assumption we always have

$$A \begin{pmatrix} \ell_1, \ldots, \ell_k \\ j+1, \ldots, j+k \end{pmatrix} \ge 0,$$

and as A is nonsingular there exist $1 \le \ell_1 < \cdots < \ell_k \le n$ for which

$$A \begin{pmatrix} \ell_1, \ldots, \ell_k \\ j+1, \ldots, j+k \end{pmatrix} > 0.$$

Furthermore $Q(q)$ is strictly totally positive. Thus

$$A_q \begin{pmatrix} i+1, \ldots, i+k \\ j+1, \ldots, j+k \end{pmatrix} > 0,$$

for all $i, j \in \{0, \ldots, n-k\}$. From Theorem 2.2 A_q is strictly totally positive. In addition,

$$\lim_{q \to 0^+} A_q = A,$$

and therefore A is totally positive. □

Theorems 2.2 and 2.3 are very useful tools for proving strict total positivity. We now settle a previous debt by presenting proofs of Propositions 1.9 and 1.10 from Section 1.2. We recall the following.

Proposition 1.9 *Assume A is an $n \times m$ strictly totally positive matrix. For given $1 \le r < n$ set*

$$c_{ij} = \begin{cases} a_{ij}, & i = 1, \ldots, r, \ j = 2, \ldots, m \\ A\begin{pmatrix} i-1, i \\ 1, j \end{pmatrix}, & i = r+1, \ldots, n, \ j = 2, \ldots, m. \end{cases}$$

Then $C = (c_{ij})$ is an $n \times (m-1)$ strictly totally positive matrix. If A is totally positive, then C is totally positive.

Proof Let $C = (c_{ij})$ be as above. We prove our desired result by obtaining an explicit formula for the arbitrary minors of C composed of consecutive rows, i.e.,

$$C\begin{pmatrix} i+1, \ldots, i+k \\ j_1, \ldots, j_k \end{pmatrix}.$$

The first statement in Proposition 1.9 will then follow from either Theorem 2.2 or 2.3.

Obviously if $i + k \le r$, then

$$C\begin{pmatrix} i+1, \ldots, i+k \\ j_1, \ldots, j_k \end{pmatrix} = A\begin{pmatrix} i+1, \ldots, i+k \\ j_1, \ldots, j_k \end{pmatrix}.$$

We prove that for $i + 1 \le r < i + k$

$$C\begin{pmatrix} i+1, \ldots, i+k \\ j_1, \ldots, j_k \end{pmatrix} = a_{r,1} \cdots a_{i+k-1,1} A\begin{pmatrix} i+1, \ldots, i+k \\ j_1, \ldots, j_k \end{pmatrix}$$

and for $r < i + 1$

$$C\begin{pmatrix} i+1, \ldots, i+k \\ j_1, \ldots, j_k \end{pmatrix} = a_{i+1,1} a_{i+2,1} \cdots a_{i+k-1,1} A\begin{pmatrix} i, i+1, \ldots, i+k \\ 1, j_1, \ldots, j_k \end{pmatrix}.$$

The first equality can be explained as follows.

$$C\begin{pmatrix} i+1,\ldots,i+k \\ j_1,\ldots,j_k \end{pmatrix} = \begin{vmatrix} a_{i+1,j_1} & \cdots & a_{i+1,j_k} \\ \vdots & & \vdots \\ a_{r,j_1} & \cdots & a_{r,j_k} \\ A\binom{r,r+1}{1,j_1} & \cdots & A\binom{r,r+1}{1,j_k} \\ \vdots & & \vdots \\ A\binom{i+k-1,i+k}{1,j_1} & \cdots & A\binom{i+k-1,i+k}{1,j_k} \end{vmatrix}.$$

By adding $a_{r+1,1}$ times the $(r-i)$th row (which is the row with entries $(a_{r,j_1},\ldots,a_{r,j_k})$) to the succeeding row, we get a new $(r-i+1)$st row of

$$(a_{r,1}a_{r+1,j_1},\ldots,a_{r,1}a_{r+1,j_k}).$$

Factor out $a_{r,1}$ and add $a_{r+2,1}$ times this row to the next row to obtain a new $(r-i+2)$nd row of

$$(a_{r+1,1}a_{r+2,j_1},\ldots,a_{r+1,1}a_{r+2,j_k}),$$

etc. The formula now easily follows.

The second equality is similarly obtained by considering the determinant

$$\begin{vmatrix} a_{i1} & a_{ij_1} & \cdots & a_{ij_k} \\ 0 & A\binom{i,i+1}{1,j_1} & \cdots & A\binom{i,i+1}{1,j_k} \\ \vdots & \vdots & & \vdots \\ 0 & A\binom{i+k-1,i+k}{1,j_1} & \cdots & A\binom{i+k-1,i+k}{1,j_k} \end{vmatrix}.$$

On the one hand this determinant equals

$$a_{i1}C\begin{pmatrix} i+1,\ldots,i+k \\ j_1,\ldots,j_k \end{pmatrix}.$$

On the other hand, adding $a_{i+1,1}$ times the first row to the second row we get a new second row of

$$(a_{i1}a_{i+1,1},a_{i1}a_{i+1,j_1},\ldots,a_{i1}a_{i+1,j_k}).$$

Factor out a_{i1} and add $a_{i+2,1}$ times the second row to the third row etc. The formula now easily follows.

From these three formulæ it follows from Theorems 2.2 or 2.3 that if A is strictly totally positive then C is strictly totally positive. Now assume that A is only totally positive. From Theorem 2.6 there exists a sequence (A_k) of $n \times m$ strictly totally positive matrices whose elements converge to the corresponding elements of A. Construct the associated C_k (in the same way in which C is constructed from A). Each C_k is strictly totally positive and the elements of C_k converge to the corresponding elements of C. Thus C is totally positive. $\square$

Proposition 1.10 *Assume A is an $n \times m$ strictly totally positive matrix. Fix $p < \min\{m, n\}$ and set*

$$c_{ij} = A\begin{pmatrix} i-p, \ldots, i-1, i \\ 1, \ldots, p, j \end{pmatrix}$$

for $i = p+1, \ldots, n$, $j = p+1, \ldots, m$. Then $C = (c_{ij})$ is an $(n-p) \times (m-p)$ strictly totally positive matrix.

Proof We prove Proposition 1.10 as a consequence of Proposition 1.9. As A is a strictly totally positive matrix, Proposition 1.10 holds for $p = 1$ since the associated C is a submatrix of the matrix C in Proposition 1.9 with $r = 1$. We apply an induction argument on p. Let us assume the validity of Proposition 1.10 for $p - 1$. That is, setting

$$e_{ij} = A\begin{pmatrix} i-p+1, \ldots, i-1, i \\ 1, \ldots, p-1, j \end{pmatrix}$$

for $i = p, \ldots, n$, $j = p, \ldots, m$, we assume that $E = (e_{ij})$ is an $(n-p+1) \times (m-p+1)$ strictly totally positive matrix. Now define

$$d_{ij} = E\begin{pmatrix} i-1, i \\ p, j \end{pmatrix}$$

for $i = p+1, \ldots, n$, $j = p+1, \ldots, m$. Applying Proposition 1.9 with $r = 1$ it follows that $D = (d_{ij})$ is an $(n-p) \times (m-p)$ strictly totally positive matrix. In addition, from Sylvester's Determinant Identity,

$$\begin{aligned} d_{ij} &= E\begin{pmatrix} i-1, i \\ p, j \end{pmatrix} = \begin{vmatrix} A\begin{pmatrix} i-p,\ldots,i-2,i-1 \\ 1,\ldots,p-1,p \end{pmatrix} & A\begin{pmatrix} i-p,\ldots,i-2,i-1 \\ 1,\ldots,p-1,j \end{pmatrix} \\ A\begin{pmatrix} i-p+1,\ldots,i-1,i \\ 1,\ldots,p-1,p \end{pmatrix} & A\begin{pmatrix} i-p+1,\ldots,i-1,i \\ 1,\ldots,p-1,j \end{pmatrix} \end{vmatrix} \\ &= A\begin{pmatrix} i-p+1, \ldots, i-1 \\ 1, \ldots, p-1 \end{pmatrix} A\begin{pmatrix} i-p, i-p+1, \ldots, i-1, i \\ 1, \ldots, p-1, p, j \end{pmatrix}. \end{aligned}$$

Note that the values

$$A\begin{pmatrix} i-p+1, \ldots, i-1 \\ 1, \ldots, p-1 \end{pmatrix}$$

are strictly positive and independent of j. Thus the matrix D being strictly totally positive is equivalent to the matrix C with elements

$$c_{ij} = A\begin{pmatrix} i-p, i-p+1, \ldots, i-1, i \\ 1, \ldots, p-1, p, j \end{pmatrix}$$

being strictly totally positive. This advances the induction step and proves the proposition. □

2.3 Triangular total positivity

A square matrix $A = (a_{ij})$ is said to be *upper triangular* if $a_{ij} = 0$ for all $i > j$. It is said to be *lower triangular* if $a_{ij} = 0$ for all $i < j$. We say that the $n \times n$ matrix $A = (a_{ij})$ is *upper strictly totally positive* if it is upper triangular and

$$A\begin{pmatrix} i_1, \ldots, i_k \\ j_1, \ldots, j_k \end{pmatrix} > 0$$

whenever $i_m \le j_m$, $m = 1, \ldots, k$. (As usual we always assume that $1 \le i_1 < \cdots < i_k \le n$ and $1 \le j_1 < \cdots < j_k \le n$.) Note that for an upper triangular matrix we always have

$$A\begin{pmatrix} i_1, \ldots, i_k \\ j_1, \ldots, j_k \end{pmatrix} = 0$$

if $i_m > j_m$ for some $m \in \{1, \ldots, k\}$. We say that the $n \times n$ matrix $A = (a_{ij})$ is *upper totally positive* if it is upper triangular and

$$A\begin{pmatrix} i_1, \ldots, i_k \\ j_1, \ldots, j_k \end{pmatrix} \ge 0$$

whenever $i_m \le j_m$, $m = 1, \ldots, k$. (This latter restriction has no meaning here.)

Similarly we say that the $n \times n$ matrix $A = (a_{ij})$ is *lower strictly totally positive* if it is lower triangular and

$$A\begin{pmatrix} i_1, \ldots, i_k \\ j_1, \ldots, j_k \end{pmatrix} > 0$$

whenever $i_m \ge j_m$, $m = 1, \ldots, k$. We say that the $n \times n$ matrix $A = (a_{ij})$ is *lower totally positive* if it is lower triangular and

$$A\begin{pmatrix} i_1, \ldots, i_k \\ j_1, \ldots, j_k \end{pmatrix} \ge 0$$

whenever $i_m \ge j_m$, $m = 1, \ldots, k$.

There are relatively simple criteria for determining when a triangular matrix is strictly totally positive in the above sense. These criteria are very reminiscent of the criteria in Theorem 2.3, and for good reason.

Theorem 2.8 *Let $A = (a_{ij})$ be an $n \times n$ upper triangular matrix satisfying*

$$A\begin{pmatrix} 1, \ldots, k \\ j+1, \ldots, j+k \end{pmatrix} > 0 \tag{2.2}$$

for $j = 0, 1, \ldots, n-k$, $k = 1, \ldots, n$. Then A is upper strictly totally positive. Similarly, if $A = (a_{ij})$ is an $n \times n$ lower triangular matrix satisfying

$$A\begin{pmatrix} i+1, \ldots, i+k \\ 1, \ldots, k \end{pmatrix} > 0$$

for $i = 0, 1, \ldots, n-k$, $k = 1, \ldots, n$, then A is lower strictly totally positive.

Proof Obviously the two claims are equivalent. We prove the first claim. That is, we show that if A is an $n \times n$ upper triangular matrix satisfying (2.2), then

$$A\begin{pmatrix} i_1, \ldots, i_k \\ j_1, \ldots, j_k \end{pmatrix} > 0$$

whenever $i_m \le j_m$, $m = 1, \ldots, k$.

From (2.2) and Fekete's Lemma 2.1, it follows that

$$A\begin{pmatrix} 1, \ldots, k \\ j_1, \ldots, j_k \end{pmatrix} > 0 \tag{2.3}$$

for all $1 \le j_1 < \cdots < j_k \le n$, and $k = 1, \ldots, n$. Since A is upper triangular and satisfies (2.3) we also have that for any $1 \le j_1 < \cdots < j_k \le n$, where $i + m \le j_m$, $m = 1, \ldots, k$,

$$A\begin{pmatrix} 1, \ldots, i \\ 1, \ldots, i \end{pmatrix} A\begin{pmatrix} i+1, \ldots, i+k \\ j_1, \ldots, j_k \end{pmatrix} = A\begin{pmatrix} 1, \ldots, i, i+1, \ldots, i+k \\ 1, \ldots, i, j_1, \ldots, j_k \end{pmatrix} > 0.$$

This implies that

$$A\begin{pmatrix} i+1, \ldots, i+k \\ j_1, \ldots, j_k \end{pmatrix} > 0 \tag{2.4}$$

for any $1 \le j_1 < \cdots < j_k \le n$, where $i + m \le j_m$, $m = 1, \ldots, k$. It thus remains to prove that we can replace consecutive rows by arbitrary rows (within the confines of the upper triangularity).

To this end we rework the method of proof of Fekete's Lemma 2.1. For a given $\mathbf{i} = (i_1, \ldots, i_k) \in I_k^n$ we recall that $d(\mathbf{i})$, the *dispersion* of $\mathbf{i}$, is given by

$$d(\mathbf{i}) := \sum_{j=2}^{k}(i_j - i_{j-1} - 1) = (i_k - i_1) - (k-1) \geq 0.$$

Then $d(\mathbf{i}) = 0$ if and only if the sequence $\mathbf{i}$ is composed of consecutive integers. We prove the result by induction on both k and on the value of $d(\cdot)$.

We assume that

$$A\begin{pmatrix} i_1, \ldots, i_s \\ j_1, \ldots, j_s \end{pmatrix} > 0$$

whenever $i_m \leq j_m$, $m = 1, \ldots, s$, and $s \leq k$ and $d(\mathbf{i}) < p$. From (2.4) the result is valid for every k when $d(\mathbf{i}) = 0$.

Let $\mathbf{i} = (i_1, \ldots, i_k)$ with $d(\mathbf{i}) = p > 0$, and $i_m \leq j_m$, $m = 1, \ldots, k$. We claim that

$$A\begin{pmatrix} i_1, \ldots, i_k \\ j_1, \ldots, j_k \end{pmatrix} > 0.$$

As the sequence $\mathbf{i}$ is not composed of consecutive integers, we can add to this sequence an integer between i_1 and i_k. Let $1 \leq r_1 < \cdots < r_{k+1} \leq n$ be the resulting sequence, assuming r_ℓ to be the new index $(1 < \ell < k+1)$. Thus

$$\begin{aligned} r_m &\leq j_m, & m &= 1, \ldots, \ell - 1 \\ r_m < r_{m+1} &\leq j_m, & m &= \ell, \ldots, k. \end{aligned} \qquad (2.5)$$

By the determinantal identity (1.2) of Section 1.1, we have

$$\begin{aligned} &A\begin{pmatrix} r_1, \ldots, \widehat{r_\ell}, \ldots, r_{k+1} \\ j_1, \ldots, j_k \end{pmatrix} A\begin{pmatrix} r_2, \ldots, r_k \\ j_2, \ldots, j_k \end{pmatrix} \\ &\quad = A\begin{pmatrix} r_2, \ldots, r_{k+1} \\ j_1, \ldots, j_k \end{pmatrix} A\begin{pmatrix} r_1, \ldots, \widehat{r_\ell}, \ldots, r_k \\ j_2, \ldots, j_k \end{pmatrix} \\ &\qquad + A\begin{pmatrix} r_1, \ldots, r_k \\ j_1, \ldots, j_k \end{pmatrix} A\begin{pmatrix} r_2, \ldots, \widehat{r_\ell}, \ldots, r_{k+1} \\ j_2, \ldots, j_k \end{pmatrix}. \end{aligned}$$

Now

$$A\begin{pmatrix} r_2, \ldots, r_k \\ j_2, \ldots, j_k \end{pmatrix} > 0$$

by the induction assumption on k and since $r_m \leq j_m$, $m = 2, \ldots, k$ (see (2.5)), while

$$A\begin{pmatrix} r_1, \ldots, r_k \\ j_1, \ldots, j_k \end{pmatrix} > 0$$

since $r_m \le j_m$, $m = 1, \ldots, k$ (see (2.5)) and by the induction assumption on $d(\mathbf{i})$. That is, since $r_1 = i_1$ and $r_k < i_k$ it follows that $d((r_1, \ldots, r_k)) < d(\mathbf{i})$. Furthermore

$$A\begin{pmatrix} r_2, \ldots, \widehat{r}_\ell, \ldots, r_{k+1} \\ j_2, \ldots, j_k \end{pmatrix} > 0$$

by the induction hypothesis on k and since $r_m \le j_m$, $m = 2, \ldots, \ell - 1$, and $r_{m+1} \le j_m$, $m = \ell, \ldots, k$ (see (2.5)). Similarly we have that

$$A\begin{pmatrix} r_1, \ldots, \widehat{r}_\ell, \ldots, r_k \\ j_2, \ldots, j_k \end{pmatrix} > 0.$$

Now

$$A\begin{pmatrix} r_2, \ldots, r_{k+1} \\ j_1, \ldots, j_k \end{pmatrix} \ge 0,$$

because this determinant vanishes if we have $r_{m+1} > j_m$ for some $m \in \{1, \ldots, \ell - 1\}$ (which is possible), while if not then it is strictly positive since $d((r_2, \ldots, r_{k+1})) < d(\mathbf{i})$.

It therefore follows that

$$A\begin{pmatrix} i_1, \ldots, i_k \\ j_1, \ldots, j_k \end{pmatrix} = A\begin{pmatrix} r_1, \ldots, \widehat{r}_\ell, \ldots, r_{k+1} \\ j_1, \ldots, j_k \end{pmatrix} > 0 \quad ,$$

which advances the induction hypothesis and proves the theorem. □

In parallel with Proposition 2.5 we have the following result, which we state for upper triangular matrices. It is also an immediate consequence of Theorem 1.19.

Proposition 2.9 *Let A be an $n \times n$ upper totally positive matrix. Then A is upper strictly totally positive if*

$$A\begin{pmatrix} 1, \ldots, k \\ n-k+1, \ldots, n \end{pmatrix} > 0$$

for $k = 1, \ldots, n$.

2.4 *LDU* factorizations

Assume $A = (a_{ij})$ is an $n \times n$ strictly totally positive matrix. Since

$$A\begin{pmatrix} 1, \ldots, k \\ 1, \ldots, k \end{pmatrix} > 0$$

for $k = 1, \ldots, n$, then, as is well known, A has a unique *LU factorization*. In fact A can be decomposed into the form

$$A = LDU,$$

where $L = (\ell_{ij})$ is a unit diagonal lower triangular matrix, $D = (d_{ij})$ is a diagonal matrix, and $U = (u_{ij})$ is a unit diagonal upper triangular matrix. The nonzero elements of L, D and U are explicitly given by

$$\begin{aligned}
\ell_{ij} &= \frac{A\begin{pmatrix}1,\ldots,j-1,i\\1,\ldots,j-1,j\end{pmatrix}}{A\begin{pmatrix}1,\ldots,j-1,j\\1,\ldots,j-1,j\end{pmatrix}}, \qquad i \geq j,\\
u_{ij} &= \frac{A\begin{pmatrix}1,\ldots,i-1,i\\1,\ldots,i-1,j\end{pmatrix}}{A\begin{pmatrix}1,\ldots,i-1,i\\1,\ldots,i-1,i\end{pmatrix}}, \qquad i \leq j,\\
d_{ii} &= \frac{A\begin{pmatrix}1,\ldots,i-1,i\\1,\ldots,i-1,i\end{pmatrix}}{A\begin{pmatrix}1,\ldots,i-1\\1,\ldots,i-1\end{pmatrix}}.
\end{aligned}$$

Using Theorem 2.8 we prove that L is a lower strictly totally positive matrix and U is an upper strictly totally positive matrix.

Theorem 2.10 *Assume A is a square strictly totally positive matrix. Then A has a unique factorization of the form*

$$A = LDU,$$

where L is a unit diagonal lower strictly totally positive matrix, U is a unit diagonal upper strictly totally positive matrix, and D is a diagonal matrix whose diagonal entries are strictly positive.

Proof We need only prove that L is a lower strictly totally positive matrix and U is an upper strictly totally positive matrix. The other facts follow trivially. As $A = LDU$ and D is a diagonal matrix, then from the Cauchy–Binet formula

$$\begin{aligned}
&A\begin{pmatrix}1,\ldots,k\\j+1,\ldots,j+k\end{pmatrix}\\
&\quad= \sum_{1\leq\ell_1<\cdots<\ell_k\leq n} L\begin{pmatrix}1,\ldots,k\\\ell_1,\ldots,\ell_k\end{pmatrix} D\begin{pmatrix}\ell_1,\ldots,\ell_k\\\ell_1,\ldots,\ell_k\end{pmatrix} U\begin{pmatrix}\ell_1,\ldots,\ell_k\\j+1,\ldots,j+k\end{pmatrix}.
\end{aligned}$$

Since L is lower triangular

$$L\begin{pmatrix}1,\ldots,k\\ \ell_1,\ldots,\ell_k\end{pmatrix}=0$$

except when $\ell_m=m$, $m=1,\ldots,k$. Thus

$$A\begin{pmatrix}1,\ldots,k\\ j+1,\ldots,j+k\end{pmatrix}=L\begin{pmatrix}1,\ldots,k\\ 1,\ldots,k\end{pmatrix}D\begin{pmatrix}1,\ldots,k\\ 1,\ldots,k\end{pmatrix}U\begin{pmatrix}1,\ldots,k\\ j+1,\ldots,j+k\end{pmatrix}.$$

Now

$$L\begin{pmatrix}1,\ldots,k\\ 1,\ldots,k\end{pmatrix}=1,\quad D\begin{pmatrix}1,\ldots,k\\ 1,\ldots,k\end{pmatrix}>0,\quad\text{and}\quad A\begin{pmatrix}1,\ldots,k\\ j+1,\ldots,j+k\end{pmatrix}>0.$$

Therefore

$$U\begin{pmatrix}1,\ldots,k\\ j+1,\ldots,j+k\end{pmatrix}>0.$$

As this is valid for all appropriate j and k we have from Theorem 2.8 that U is an upper strictly totally positive matrix. The analogous proof shows that L is a lower strictly totally positive matrix. □

Each square strictly totally positive matrix also has a unique factorization of the form

$$A=UDL,$$

where U is a unit diagonal upper strictly totally positive matrix, L is a unit diagonal lower strictly totally positive matrix, and D is a diagonal matrix whose diagonal entries are strictly positive. The nonzero elements of U, D, and L are explicitly given by

$$\begin{aligned}
u_{ij} &= \frac{A\begin{pmatrix}i,j+1,\ldots,n\\ j,j+1,\ldots,n\end{pmatrix}}{A\begin{pmatrix}j,j+1,\ldots,n\\ j,j+1,\ldots,n\end{pmatrix}}, && i\le j,\\
\ell_{ij} &= \frac{A\begin{pmatrix}i,i+1,\ldots,n\\ j,i+1,\ldots,n\end{pmatrix}}{A\begin{pmatrix}i,i+1,\ldots,n\\ i,i+1,\ldots,n\end{pmatrix}}, && i\ge j,\\
d_{ii} &= \frac{A\begin{pmatrix}i,i+1,\ldots,n\\ i,i+1,\ldots,n\end{pmatrix}}{A\begin{pmatrix}i+1,\ldots,n\\ i+1,\ldots,n\end{pmatrix}}.
\end{aligned}$$

This can be shown in a variety of ways. It also follows easily from the previous result and the following. Let

$$Q = \begin{pmatrix} 0 & \cdots & 0 & 1 \\ 0 & \cdots & 1 & 0 \\ \vdots & & \vdots & \vdots \\ 1 & \cdots & 0 & 0 \end{pmatrix}.$$

Then

$$QAQ$$

is the matrix obtained from A by reversing the order of its rows and columns. As is known (see Proposition 1.3 in Chapter 1), A is strictly totally positive if and only if QAQ is strictly totally positive. From the LDU factorization of QAQ (Theorem 2.10) we have

$$A = (QLQ)(QDQ)(QUQ).$$

QLQ is a unit diagonal upper strictly totally positive matrix while QUQ is a unit diagonal lower strictly totally positive matrix.

Another consequence of Theorem 2.10 is an alternative, more elementary, proof of Theorem 2.3.

Another Proof of Theorem 2.3. Since

$$A\begin{pmatrix} 1,\ldots,k \\ 1,\ldots,k \end{pmatrix} > 0$$

for $k = 1,\ldots,n$, it follows that A has a unique LDU factorization. Since

$$A\begin{pmatrix} 1,\ldots,k \\ j+1,\ldots,j+k \end{pmatrix}, \, A\begin{pmatrix} i+1,\ldots,i+k \\ 1,\ldots,k \end{pmatrix} > 0$$

for all appropriate j, i, and k, it follows from the proof of Theorem 2.10 that L is a unit diagonal lower strictly totally positive matrix, U is a unit diagonal upper strictly totally positive matrix, while D is a diagonal matrix whose diagonal entries are strictly positive. From this it easily follows that $A = LDU$ is strictly totally positive. □

What about totally positive matrices? Do they always have a similar unique LDU factorization? If A is also non-singular, then the answer is yes. The proof is essentially the same as the above proof, except that L is a unit diagonal lower totally positive matrix and U is a unit diagonal upper totally positive matrix. Recall that since A is totally positive and nonsingular then all principal minors of A are strictly positive (Theorem 1.13). We state this result here for easy reference.

Proposition 2.11 *Assume A is a square nonsingular totally positive matrix. Then A has a unique factorization of the form*

$$A = LDU$$

where L is a unit diagonal lower totally positive matrix, U is a unit diagonal upper totally positive matrix, and D is a diagonal matrix whose diagonal entries are strictly positive.

However, if A is singular then we must modify our demands somewhat. In general we lose the unit triangular nature of the L and U and the uniqueness. What we do have is the following.

Theorem 2.12 *Assume A is a square totally positive matrix. Then A has a factorization of the form*

$$A = LU$$

where L is a lower totally positive matrix and U is an upper totally positive matrix.

Proof From Theorem 2.6 there exists a sequence of strictly totally positive matrices (A_m) that approximate A, i.e., such that

$$\lim_{m \to \infty} A_m = A.$$

From Theorem 2.10, each $A_m = (a_{ij}(m))$ has a unique factorization of the form

$$A_m = L_m D_m U_m$$

where L_m is a unit diagonal lower strictly totally positive matrix, U_m is a unit diagonal upper strictly totally positive matrix, and D_m is a diagonal matrix whose diagonal entries are strictly positive.

There are many different normalizations to this factorization that allow us to take limits. Here is one such normalization. By pre- and postmultiplying by positive diagonal matrices we can factor A_m in the form

$$A_m = \widetilde{D}_m \widetilde{L}_m \widetilde{U}_m$$

where $\widetilde{L}_m$ and $\widetilde{U}_m$ are stochastic (rows sums 1). As a consequence we also have

$$\sum_j a_{ij}(m) = \widetilde{d}_{ii}(m)$$

for all i, where $\widetilde{d}_{ii}(m)$ is the (i,i) entry of $\widetilde{D}_m$. Thus all entries in each of the factors are uniformly bounded. Now that we have bounds on the elements of $\widetilde{D}_m$, $\widetilde{L}_m$ and $\widetilde{U}_m$, we may take a subsequence along which each of the entries of these matrices converges. Let their respective limits be $\widehat{D}$, $\widehat{L}$ and U, and set $L = \widehat{D}\widehat{L}$. Then

$$A = LU.$$

Furthermore it follows that L is a lower totally positive matrix and U is an upper totally positive matrix. □

2.5 Criteria for total positivity

Unfortunately, nothing quite as simple as the conditions in Theorems 2.2 and 2.3 seem to be valid for determining if a matrix is totally positive.

It is possible, for example, that every minor composed from consecutive rows and consecutive columns is nonnegative and yet A is not totally positive. A simple example is the 4×4 matrix

$$A \begin{pmatrix} 1 & 0 & 0 & 0 \\ 0 & 1 & 0 & 0 \\ 0 & 0 & 1 & 0 \\ 1 & 0 & 0 & 1 \end{pmatrix}.$$

One therefore needs somewhat more restrictive criteria.

Let us again recall that for a given strictly increasing sequence $\mathbf{i} = (i_1, \ldots, i_k) \in I_k^n$ we define $d(\mathbf{i})$, the *dispersion* of $\mathbf{i}$, by

$$d(\mathbf{i}) := \sum_{j=2}^{k} (i_j - i_{j-1} - 1) = (i_k - i_1) - (k-1) \geq 0.$$

Note that $d(\mathbf{i})$ counts the number of integers between i_1 and i_k that are not in the sequence $i_1, \ldots, i_k$, and thus $d(\mathbf{i}) = 0$ if and only if the sequence $\mathbf{i}$ is composed of consecutive integers.

The main theorem providing determinantal criteria for when a matrix is totally positive is the following. It is based on previous reasonings and techniques, but is technically more detailed.

Theorem 2.13 *An $n \times m$ matrix $A = (a_{ij})$ of rank r is totally positive if*

$$A\begin{pmatrix} i_1, \ldots, i_k \\ j_1, \ldots, j_k \end{pmatrix} \geq 0 \tag{2.6}$$

for all $\mathbf{i} = (i_1, \ldots, i_k)$ satisfying $d(\mathbf{i}) \leq n - r$, all $\mathbf{j} = (j_1, \ldots, j_k)$ and $k = 1, \ldots, r$.

Note that if A is nonsingular, i.e., $r = n = m$, then this is Proposition 2.7.

Proof We wish to prove that from (2.6) it follows that

$$A\begin{pmatrix} i_1, \ldots, i_s \\ j_1, \ldots, j_s \end{pmatrix} \geq 0 \tag{2.7}$$

for all $1 \leq i_1 < \cdots < i_s \leq n$ and $1 \leq j_1 < \cdots < j_s \leq m$ and all s.

The proof will be based on an induction argument. The induction is on s. For $s = 1$ there is nothing to prove since $d(\{i_1\}) = 0 \leq n - r$ and thus (2.7) follows from (2.6). Let us suppose that (2.7) holds for all $s \leq k - 1$, but is not valid for some $(i_1, \ldots, i_k)$ and $(j_1, \ldots, j_k)$, as above. That is, we have

$$A\begin{pmatrix} i_1, \ldots, i_k \\ j_1, \ldots, j_k \end{pmatrix} < 0. \tag{2.8}$$

By our assumption, we must have $d(\mathbf{i}) > n - r$. Choose $\mathbf{i}^* = (i_1^*, \ldots, i_k^*)$ and $\mathbf{j}^* = (j_1^*, \ldots, j_k^*)$ such that (2.8) holds and $d(\mathbf{i}^*) = \ell$ is minimal, i.e., for given k and every $\mathbf{i} \in I_k^n$, $\mathbf{j} \in I_k^m$ satisfying (2.8) we have $d(\mathbf{i}) \geq \ell$. If $\mathbf{a}^i$ denotes the ith row of A, then we prove that for all p satisfying $i_1^* < p < i_k^*$ we have that $\mathbf{a}^p \in \operatorname{span}\{\mathbf{a}^{i_2^*}, \ldots, \mathbf{a}^{i_{k-1}^*}\}$. As there are exactly ℓ such p (since $d(\mathbf{i}^*) = \ell$), it follows that the rank of A is at most $n - \ell$. But, by assumption, $\ell > n - r$, which implies that the rank of A is strictly less than r. A contradiction.

It therefore remains to prove that for all p satisfying $i_1^* < p < i_k^*$ we have $\mathbf{a}^p \in \operatorname{span}\{\mathbf{a}^{i_2^*}, \ldots, \mathbf{a}^{i_{k-1}^*}\}$. This we prove in two steps. We first prove that it is true when we restrict ourselves to the columns $j_1^*, \ldots, j_k^*$. We then subsequently extend it to all columns.

Let p satisfy $i_1^* < p < i_k^*$, and let $r_1 < \cdots < r_{k+1}$ be the sequence obtained by adding p to the sequence $i_1^*, \ldots, i_k^*$. Assume $p = r_t$. From the determinantal equality (1.2) of Section 1.1, we have for each $s \in \{1, \ldots, k\}$:

$$\begin{aligned} A\begin{pmatrix} r_1, \ldots, \widehat{r}_t, \ldots, r_{k+1} \\ j_1^*, \ldots, j_k^* \end{pmatrix} A\begin{pmatrix} r_2, \ldots, r_k \\ j_1^*, \ldots, \widehat{j}_s^*, \ldots, j_k^* \end{pmatrix} \\ = A\begin{pmatrix} r_2, \ldots, r_{k+1} \\ j_1^*, \ldots, j_k^* \end{pmatrix} A\begin{pmatrix} r_1, \ldots, \widehat{r}_t, \ldots, r_k \\ j_1^*, \ldots, \widehat{j}_s^*, \ldots, j_k^* \end{pmatrix} \\ + A\begin{pmatrix} r_1, \ldots, r_k \\ j_1^*, \ldots, j_k^* \end{pmatrix} A\begin{pmatrix} r_2, \ldots, \widehat{r}_t, \ldots, r_{k+1} \\ j_1^*, \ldots, \widehat{j}_s^*, \ldots, j_k^* \end{pmatrix}. \end{aligned} \tag{2.9}$$

By the induction assumption, all three $k-1 \times k-1$ minors in (2.9) are nonnegative, while the two $k \times k$ minors on the right-hand-side of the equation are nonnegative since the dispersion of the rows therein is strictly less than ℓ, and the minimality property of ℓ. Thus the value of the right-hand-side of (2.9) is nonnegative. As (2.8) holds for $i_1^*, \ldots, i_k^*$ and $j_1^*, \ldots, j_k^*$ (which is the left-most determinant in (2.9)), it necessarily follows that we must have

$$A\begin{pmatrix} i_2^*, \ldots, p, \ldots, i_{k-1}^* \\ j_1^*, \ldots, \widehat{j}_s^*, \ldots, j_k^* \end{pmatrix} = A\begin{pmatrix} r_2, \ldots, r_k \\ j_1^*, \ldots, \widehat{j}_s^*, \ldots, j_k^* \end{pmatrix} = 0. \tag{2.10}$$

As (2.10) is valid for all $s = 1, \ldots, k$, we have $\mathbf{a}^p \in \operatorname{span}\{\mathbf{a}^{i_2^*}, \ldots, \mathbf{a}^{i_{k-1}^*}\}$ on the columns $j_1^*, \ldots, j_k^*$, for each p satisfying $i_1^* < p < i_k^*$.

Since

$$A\begin{pmatrix} i_1^*, \ldots, i_k^* \\ j_1^*, \ldots, j_k^* \end{pmatrix} < 0,$$

the rows $\mathbf{a}^{i_1^*}, \ldots, \mathbf{a}^{i_k^*}$ are linearly independent on the columns $j_1^*, \ldots, j_k^*$. Thus the rows $\mathbf{a}^{i_2^*}, \ldots, \mathbf{a}^{i_{k-1}^*}$ are linearly independent on some columns $\widetilde{j}_1, \ldots, \widetilde{j}_{k-2}$, where $\{\widetilde{j}_1, \ldots, \widetilde{j}_{k-2}\} = \{j_1^*, \ldots, j_k^*\} \backslash \{j_s^*, j_t^*\}$ (we assume that $j_s^* < j_t^*$ in what follows). That is,

$$A\begin{pmatrix} i_2^*, \ldots, i_{k-1}^* \\ \widetilde{j}_1, \ldots, \widetilde{j}_{k-2} \end{pmatrix} \neq 0.$$

By our induction assumption, this implies that the above minor is, in fact, strictly positive. We use

$$A\begin{bmatrix} i_2^*, \ldots, i_{k-1}^* \\ \widetilde{j}_1, \ldots, \widetilde{j}_{k-2} \end{bmatrix}$$

as a pivot block in Sylvester's Determinant Identity. That is, set

$$d_{ij} = A\begin{pmatrix} i_2^*, \ldots, i_{k-1}^*, i \\ \widetilde{j}_1, \ldots, \widetilde{j}_{k-2}, j \end{pmatrix}.$$

Then by our induction hypothesis on k we have $d_{ij} \geq 0$ for all i and j, and from (2.10) we have $d_{pj_s^*} = d_{pj_t^*} = 0$ for each p between i_1^* and i_k^*. Furthermore, by Sylvester's Determinant Identity we have

$$D\begin{pmatrix} i_1^*, i_k^* \\ j_s^*, j_t^* \end{pmatrix} = A\begin{pmatrix} i_2^*, \ldots, i_{k-1}^* \\ \tilde{j}_1, \ldots, \tilde{j}_{k-2} \end{pmatrix} A\begin{pmatrix} i_1^*, \ldots, i_k^* \\ j_1^*, \ldots, j_k^* \end{pmatrix} < 0.$$

Thus we must have $d_{i_1^* j_t^*} > 0$ and $d_{i_k^* j_s^*} > 0$.

From Sylvester's Determinant Identity, for $j < j_t^*$,

$$D\begin{pmatrix} i_1^*, p \\ j, j_t^* \end{pmatrix} = A\begin{pmatrix} i_2^*, \ldots, i_{k-1}^* \\ \tilde{j}_1, \ldots, \tilde{j}_{k-2} \end{pmatrix} A\begin{pmatrix} i_1^*, \ldots, i_{k-1}^*, p \\ \tilde{j}_1, \ldots, \tilde{j}_{k-2}, j, j_t^* \end{pmatrix}.$$

Both terms of the right-hand side of this equality are nonnegative. The first of these two terms is strictly positive. The second term is nonnegative because the dispersion of $i_1^*, \ldots, i_{k-1}^*, p$ is strictly less than ℓ. Thus

$$0 \leq D\begin{pmatrix} i_1^*, p \\ j, j_t^* \end{pmatrix} = d_{i_1^* j} d_{pj_t^*} - d_{i_1^* j_t^*} d_{pj}.$$

Now $d_{pj_t^*} = 0$, $d_{i_1^* j_t^*} > 0$, and $d_{pj} \geq 0$. Thus $d_{pj} = 0$ for all $j < j_t^*$. When $j > j_s^*$ we consider

$$D\begin{pmatrix} p, i_k^* \\ j_s^*, j \end{pmatrix}$$

and apply the same reasoning to obtain $d_{pj} = 0$ for all $j > j_s^*$. Thus for all j we have

$$0 = d_{pj} = A\begin{pmatrix} i_2^*, \ldots, i_{k-1}^*, p \\ \tilde{j}_1, \ldots, \tilde{j}_{k-2}, j \end{pmatrix}.$$

Since

$$A\begin{pmatrix} i_2^*, \ldots, i_{k-1}^* \\ \tilde{j}_1, \ldots, \tilde{j}_{k-2} \end{pmatrix} > 0,$$

this implies that row $\mathbf{a}^p$ is in span$\{\mathbf{a}^{i_2^*}, \ldots, \mathbf{a}^{i_{k-1}^*}\}$ over all columns of A, and this is true for all p satisfying $i_1^* < p < i_k^*$. □

One fairly elementary consequence of the above is the following result.

Theorem 2.14 *Let $A = (a_{ij})$ be an $n \times n$ nonsingular upper triangular matrix satisfying*

$$A\begin{pmatrix} 1, \ldots, k \\ j_1, \ldots, j_k \end{pmatrix} \geq 0$$

for all $1 \leq j_1 < \cdots < j_k \leq n$, $k = 1, \ldots, n$. Then A is upper totally *positive. Similarly, if $A = (a_{ij})$ is an $n \times n$ nonsingular lower triangular matrix satisfying*

$$A\begin{pmatrix} i_1, \ldots, i_k \\ 1, \ldots, k \end{pmatrix} \geq 0$$

for $1 \leq i_1 < \cdots < i_k \leq n$, $k = 1, \ldots, n$, then A is lower totally positive.

Proof We prove the first claim. As A is a nonsingular upper triangular matrix we necessarily have $a_{mm} > 0$ for $m = 1, \ldots, n$. From Theorem 2.13 it suffices to prove that

$$A\begin{pmatrix} i+1, \ldots, i+k \\ j_1, \ldots, j_k \end{pmatrix} \geq 0$$

for all appropriate i, $1 \leq j_1 < \cdots < j_k \leq n$ and k (i.e., we must check the case $d(\mathbf{i}) = 0$). It suffices to consider only those minors with $i + m \leq j_m$, $m = 1, \ldots, k$, since all other minors vanish due to the upper triangular structure of A. Now, by assumption,

$$0 \leq A\begin{pmatrix} 1, \ldots, i, i+1, \ldots, i+k \\ 1, \ldots, i, j_1, \ldots, j_k \end{pmatrix} = \prod_{m=1}^{i} a_{mm}\, A\begin{pmatrix} i+1, \ldots, i+k \\ j_1, \ldots, j_k \end{pmatrix}.$$

As $a_{mm} > 0$ for all m, this proves our result. □

Based on the above Theorem 2.14 and the analysis of Section 2.4 we have the following result, which should be compared to Theorem 2.13 and Proposition 2.7.

Proposition 2.15 *Let $A = (a_{ij})$ be an $n \times n$ nonsingular matrix. Then A is totally positive if and only if A satisfies the following:*

(i) $A\begin{pmatrix} 1, \ldots, k \\ 1, \ldots, k \end{pmatrix} > 0, \qquad k = 1, \ldots, n$

(ii) $A\begin{pmatrix} i_1, \ldots, i_k \\ 1, \ldots, k \end{pmatrix} \geq 0, \qquad 1 \leq i_1 < \cdots < i_k \leq n,\ k = 1, \ldots, n$

(iii) $A\begin{pmatrix} 1, \ldots, k \\ j_1, \ldots, j_k \end{pmatrix} \geq 0, \qquad 1 \leq j_1 < \cdots < j_k \leq n,\ k = 1, \ldots, n.$

Proof If A is totally positive then (ii) and (iii) hold. If A is also nonsingular then (i) holds from Theorem 1.13.

Assume (i) – (iii) hold. The LDU factorization of A, as in Section 2.4, is well defined since (i) holds. From (i), (ii), and (iii) we see that D is a diagonal matrix whose diagonal entries are strictly positive, L is a unit diagonal lower triangular matrix, and U is a unit diagonal upper triangular matrix. Now, from the proof of Theorem 2.10 we have,

$$A\begin{pmatrix} i_1,\ldots,i_k \\ 1,\ldots,k \end{pmatrix} = \sum_{1\le \ell_1<\cdots<\ell_k\le n} L\begin{pmatrix} i_1,\ldots,i_k \\ \ell_1,\ldots,\ell_k \end{pmatrix} D\begin{pmatrix} \ell_1,\ldots,\ell_k \\ \ell_1,\ldots,\ell_k \end{pmatrix} U\begin{pmatrix} \ell_1,\ldots,\ell_k \\ 1,\ldots,k \end{pmatrix}$$

$$= L\begin{pmatrix} i_1,\ldots,i_k \\ 1,\ldots,k \end{pmatrix} D\begin{pmatrix} 1,\ldots,k \\ 1,\ldots,k \end{pmatrix}$$

and similarly

$$A\begin{pmatrix} 1,\ldots,k \\ j_1,\ldots,j_k \end{pmatrix} = D\begin{pmatrix} 1,\ldots,k \\ 1,\ldots,k \end{pmatrix} U\begin{pmatrix} 1,\ldots,k \\ j_1,\ldots,j_k \end{pmatrix}.$$

Since

$$D\begin{pmatrix} 1,\ldots,k \\ 1,\ldots,k \end{pmatrix} > 0$$

we have as a consequence of (ii) and (iii)

$$L\begin{pmatrix} i_1,\ldots,i_k \\ 1,\ldots,k \end{pmatrix},\ U\begin{pmatrix} 1,\ldots,k \\ j_1,\ldots,j_k \end{pmatrix} \ge 0.$$

This is valid for all $1 \le i_1 < \cdots < i_k \le n$, $1 \le j_1 < \cdots < j_k \le n$, and $k = 1,\ldots,n$. Thus from Theorem 2.14 L is a lower totally positive matrix and U is an upper totally positive matrix, implying in turn that

$$A = LDU$$

is totally positive. □

2.6 "Simple" criteria for strict total positivity

There is a determinantal criterion, very much different from those of the previous sections, that suffices for total positivity. It is only sufficient and not, of course, necessary. It is based solely on the sign of the elements of A and a form of measure of the magnitude of the 2×2 minors of A with consecutive rows and columns. It is both a surprising and a beautiful result. But its proof is far from trivial. It is the following.

Theorem 2.16 *Let $A = (a_{ij})$ be an $n \times n$ matrix whose entries are strictly positive. Assume*

$$a_{ij}a_{i+1,j+1} > \left(4\cos^2\frac{\pi}{n+1}\right) a_{i,j+1}a_{i+1,j}$$

for all $i,j = 1,\ldots,n-1$. Then A is strictly totally positive. Furthermore the above constant $4\cos^2\frac{\pi}{n+1}$ is best possible.

By a simple perturbation argument and the denseness of strictly totally positive matrices in the class of all totally positive matrices it follows that if the inequality in Theorem 2.16 holds, but is not always strict, then A is totally positive.

An immediate consequence of this theorem, independent of n and m, is the more easily stated.

Corollary 2.17 *Let $A = (a_{ij})$ be an $n \times m$ matrix whose entries are strictly positive. Assume*

$$a_{ij}a_{i+1,j+1} \geq 4a_{i,j+1}a_{i+1,j}$$

for all $i = 1, \ldots, n-1$, $j = 1, \ldots, m-1$. Then A is strictly totally positive.

We start by proving that the constant $4\cos^2 \frac{\pi}{n+1}$ in Theorem 2.16 is best possible.

Proposition 2.18 *For any $c < 4\cos^2 \frac{\pi}{n+1}$ there exists an $n \times n$ matrix $A = (a_{ij})$ with all entries strictly positive for which*

$$a_{ij}a_{i+1,j+1} > c\, a_{i,j+1}a_{i+1,j}, \qquad i, j = 1, \ldots, n-1,$$

and yet A is **not** *totally positive.*

Proof Consider the $n \times n$ symmetric Toeplitz matrix

$$A_n(\theta) = \begin{pmatrix} 2\cos\theta & 1 & 0 & \cdots & 0 \\ 1 & 2\cos\theta & 1 & \cdots & 0 \\ 0 & 1 & 2\cos\theta & \cdots & 0 \\ \vdots & \vdots & \vdots & \ddots & \vdots \\ 0 & 0 & 0 & \cdots & 2\cos\theta \end{pmatrix},$$

where $0 \leq \theta < \pi/2$. Expanding by the last row we see that $\det A_n(\theta)$ satisfies the recurrence relation

$$\det A_n(\theta) = 2\cos\theta \det A_{n-1}(\theta) - \det A_{n-2}(\theta).$$

Now

$$\det A_1(\theta) = 2\cos\theta = \frac{\sin 2\theta}{\sin\theta},$$

and

$$\det A_2(\theta) = 4\cos^2\theta - 1 = \frac{\sin 3\theta}{\sin\theta}.$$

From the fact that

$$\sin(n+1)\theta + \sin(n-1)\theta = 2\cos\theta \sin n\theta$$

it immediately follows that

$$\det A_n(\theta) = \frac{\sin(n+1)\theta}{\sin\theta},$$

implying that for $\theta \in (\pi/(n+1), 2\pi/(n+1))$ we have

$$\det A_n(\theta) < 0.$$

For $A_n(\theta) = (a_{ij})$ we have

$$a_{ij}a_{i+1,j+1} \geq 4\cos^2\theta \, a_{i,j+1}a_{i+1,j}$$

for $i, j = 1, \ldots, n-1$. Now for any $c < 4\cos^2 \frac{\pi}{n+1}$, let

$$\theta \in (\pi/(n+1), 2\pi/(n+1))$$

be such that $c < 4\cos^2\theta$. Thus

$$a_{ij}a_{i+1,j+1} \geq c \, a_{i,j+1}a_{i+1,j}$$

for all i, j and yet

$$\det A_n(\theta) < 0,$$

i.e., $A_n(\theta)$ is far from being totally positive.

The entries of $A_n(\theta)$ are not all strictly positive. But it is not a problem to perturb $A_n(\theta)$ so that its entries are all strictly positive, it satisfies

$$a_{ij}a_{i+1,j+1} > c \, a_{i,j+1}a_{i+1,j}$$

for all $i, j = 1, \ldots, n-1$, and the perturbed $A_n(\theta)$ is not totally positive. In fact we can also do so while maintaining the Toeplitz character of the matrix. For example, keep the diagonal and first off-diagonal entries of $A_n(\theta)$ fixed and perturb all elements of the kth off-diagonal to $\varepsilon^{(k-1)^2}$ for $k = 2, \ldots, n-1$. For $\varepsilon > 0$, sufficiently small, this new matrix satisfies all our requirements. □

The proof of the main claim of Theorem 2.16 is lengthy and complicated. We therefore divide it into a series of lemmas. We start with the following.

Lemma 2.19 *Let $A = (a_{ij})$ be an $m \times m$ matrix with strictly positive entries. Assume*

$$a_{ij}a_{i+1,j+1} > c \, a_{i,j+1}a_{i+1,j}$$

for all $i, j = 1, \ldots, m-1$. *Then for* $k, \ell \geq 1$

$$a_{ij}a_{i+k,j+\ell} > c^{k\ell}a_{i,j+\ell}a_{i+k,j},$$

$i = 1, \ldots, m-k$, $j = 1, \ldots, m-\ell$.

Proof Multiply together the $k\ell$ inequalities

$$a_{i+r,j+s}a_{i+r+1,j+s+1} > c\,a_{i+r,j+s+1}a_{i+r+1,j+s}$$

for $r = 0, 1, \ldots, k-1$ and $s = 0, 1, \ldots, \ell-1$. □

Lemma 2.20 *The following are equivalent:*

(i) $F_m(c) = \sum_{j=0}^{\lfloor m/2 \rfloor} \binom{m-j}{j}(-1)^j \frac{1}{c^j}, \qquad m = 0, 1, 2, \ldots$

(ii) $F_0(c) = F_1(c) = 1$
$F_m(c) = F_{m-1}(c) - \frac{1}{c}F_{m-2}(c), \qquad m = 2, 3, \ldots$

where $\lfloor x \rfloor$ *denotes the integral part of* x.

It is a simple matter to verify the above lemma. Its proof is left to the reader.

Lemma 2.21 *Let* $F_m(c)$ *be as above. Then*

(i) *For* $c = 4\cos^2\theta$,

$$F_m(c) = \frac{\sin(m+1)\theta}{c^{m/2}\sin\theta}.$$

(ii) *For* $c_k = 4\cos^2\frac{\pi}{k+1}$,

$$F_{m-1}(c_k) - \frac{1}{c_k^2}F_{m-2}(c_k) - \frac{1}{c_k^m} \geq F_m(c_k)$$

for $k \geq 3$, $m = 2, 3, \ldots, k$.

Proof (i) follows from the trigonometric identity

$$\frac{\sin(m+1)\theta}{\sin\theta} = \sum_{j=0}^{\lfloor m/2 \rfloor}\binom{m-j}{j}(-1)^j(2\cos\theta)^{m-2j}.$$

To prove (ii), note that, from (ii) of Lemma 2.20,

$$F_{m-1}(c_k) - \frac{1}{c_k^2}F_{m-2}(c_k) - \frac{1}{c_k^m} - F_m(c_k) = \left(\frac{1}{c_k} - \frac{1}{c_k^2}\right)F_{m-2}(c_k) - \frac{1}{c_k^m}.$$

Now

$$4\cos^2\theta = \frac{\sin 3\theta}{\sin\theta} + 1.$$

Thus continuing the above and using (i) we have

$$\begin{aligned}
&= \frac{c_k - 1}{c_k^2} \frac{\sin\frac{(m-1)\pi}{k+1}}{c_k^{\frac{m-2}{2}} \sin\frac{\pi}{k+1}} - \frac{1}{c_k^m} \\
&= \frac{1}{c_k^{\frac{m+2}{2}}} \left[\frac{\sin\frac{3\pi}{k+1}}{\sin\frac{\pi}{k+1}} \frac{\sin\frac{(m-1)\pi}{k+1}}{\sin\frac{\pi}{k+1}} - \frac{1}{\left(2\cos\frac{\pi}{k+1}\right)^{m-2}} \right] \\
&\geq \frac{1}{c_k^{\frac{m+2}{2}}} \left[\frac{\sin\frac{3\pi}{k+1}}{\sin\frac{\pi}{k+1}} \frac{\sin\frac{(m-1)\pi}{k+1}}{\sin\frac{\pi}{k+1}} - 1 \right] \geq 0,
\end{aligned}$$

for $k \geq 3$ and $m = 2, 3, \ldots, k$. Looking at the last two terms we see that we actually have strict inequality except in the case $k = 3$ and $m = 2$. □

Note that from Lemma 2.21(i) we have that $F_k(c_k) = 0$ and $F_k(c_m) > 0$ for $m < k$.

The proof of Theorem 2.16 is based on an induction argument. To ease exposition we assume, in the next series of lemmas, the following.

Assumption A *Let $A = (a_{ij})$ be an $m \times m$ matrix with strictly positive entries such that*

$$a_{ij}a_{i+1,j+1} > c\, a_{i,j+1}a_{i+1,j} \tag{2.11}$$

for all $i, j = 1, \ldots, m-1$,

$$A\begin{pmatrix} r, \ldots, m \\ r, \ldots, m \end{pmatrix} > 0, \qquad r = 2, \ldots, m, \tag{2.12}$$

and

$$A\begin{pmatrix} r, \ldots, m \\ r, \ldots, m \end{pmatrix} > a_{rr} A\begin{pmatrix} r+1, \ldots, m \\ r+1, \ldots, m \end{pmatrix} - a_{r,r+1}a_{r+1,r} A\begin{pmatrix} r+2, \ldots, m \\ r+2, \ldots, m \end{pmatrix} \tag{2.13}$$

for $r = 1, \ldots, m-2$.

Lemma 2.22 *Under Assumption A*

$$A\begin{pmatrix} r, \ldots, m \\ r, \ldots, m \end{pmatrix} > a_{rr} \left[A\begin{pmatrix} r+1, \ldots, m \\ r+1, \ldots, m \end{pmatrix} - \frac{1}{c} a_{r+1,r+1} A\begin{pmatrix} r+2, \ldots, m \\ r+2, \ldots, m \end{pmatrix} \right]$$

for $r = 1, \ldots, m-2$.

Proof Using (2.13) and (2.11) with $i = j = r$

$$
\begin{aligned}
A\begin{pmatrix} r,\dots,m \\ r,\dots,m \end{pmatrix} &> a_{rr}A\begin{pmatrix} r+1,\dots,m \\ r+1,\dots,m \end{pmatrix} - a_{r,r+1}a_{r+1,r}A\begin{pmatrix} r+2,\dots,m \\ r+2,\dots,m \end{pmatrix} \\
&> a_{rr}A\begin{pmatrix} r+1,\dots,m \\ r+1,\dots,m \end{pmatrix} - \frac{1}{c}\, a_{r,r}a_{r+1,r+1}A\begin{pmatrix} r+2,\dots,m \\ r+2,\dots,m \end{pmatrix} \\
&= a_{rr}\left[A\begin{pmatrix} r+1,\dots,m \\ r+1,\dots,m \end{pmatrix} - \frac{1}{c}\, a_{r+1,r+1}A\begin{pmatrix} r+2,\dots,m \\ r+2,\dots,m \end{pmatrix}\right].
\end{aligned}
$$

□

Lemma 2.23 *Under Assumption A and if $F_k(c) > 0$, then*

$$
\begin{aligned}
&F_k(c)A\begin{pmatrix} r,\dots,m \\ r,\dots,m \end{pmatrix} - \frac{1}{c}F_{k-1}(c)a_{rr}A\begin{pmatrix} r+1,\dots,m \\ r+1,\dots,m \end{pmatrix} \\
&> a_{rr}\left[F_{k+1}(c)A\begin{pmatrix} r+1,\dots,m \\ r+1,\dots,m \end{pmatrix} - \frac{1}{c}F_k(c)a_{r+1,r+1}A\begin{pmatrix} r+2,\dots,m \\ r+2,\dots,m \end{pmatrix}\right]
\end{aligned}
$$

for $r = 1,\dots,m-2$.

Proof As $F_k(c) > 0$ we multiply the result of Lemma 2.22 by $F_k(c)$ to obtain

$$
\begin{aligned}
&F_k(c)A\begin{pmatrix} r,\dots,m \\ r,\dots,m \end{pmatrix} \\
&\quad > a_{rr}\left[F_k(c)A\begin{pmatrix} r+1,\dots,m \\ r+1,\dots,m \end{pmatrix} - \frac{1}{c}\, F_k(c)a_{r+1,r+1}A\begin{pmatrix} r+2,\dots,m \\ r+2,\dots,m \end{pmatrix}\right]
\end{aligned}
$$

for $r = 1,\dots,m-2$. Thus

$$
\begin{aligned}
F_k(c)A\begin{pmatrix} r,\dots,m \\ r,\dots,m \end{pmatrix} &- \frac{1}{c}\, F_{k-1}(c)a_{r,r}A\begin{pmatrix} r+1,\dots,m \\ r+1,\dots,m \end{pmatrix} \\
> a_{rr}\Bigg[&\left(F_k(c) - \frac{1}{c}\, F_{k-1}(c)\right)A\begin{pmatrix} r+1,\dots,m \\ r+1,\dots,m \end{pmatrix} \\
&-\frac{1}{c}\, F_k(c)a_{r+1,r+1}A\begin{pmatrix} r+2,\dots,m \\ r+2,\dots,m \end{pmatrix}\Bigg].
\end{aligned}
$$

Since

$$F_{k+1}(c) = F_k(c) - \frac{1}{c}\, F_{k-1}(c)$$

we have, continuing the above,

$$
= a_{rr}\left[F_{k+1}(c)A\begin{pmatrix} r+1,\dots,m \\ r+1,\dots,m \end{pmatrix} - \frac{1}{c}\, F_k(c)a_{r+1,r+1}A\begin{pmatrix} r+2,\dots,m \\ r+2,\dots,m \end{pmatrix}\right].
$$

□

Note the recursive nature of the inequality of Lemma 2.23. We use Lemma 2.23 in the proof of the next two lemmas.

Lemma 2.24 *Under Assumption A and if* $F_{k+\ell}(c) > 0$ *for* $\ell = 0, 1, \ldots, m-1-r$, *then*

$$F_k(c)A\begin{pmatrix} r,\ldots,m \\ r,\ldots,m \end{pmatrix} - \frac{1}{c}F_{k-1}(c)a_{r,r}A\begin{pmatrix} r+1,\ldots,m \\ r+1,\ldots,m \end{pmatrix}$$
$$> a_{rr}a_{r+1,r+1}\cdots a_{mm}F_{k+m-r+1}(c)$$

for $r = 1, \ldots, m-2$.

Proof We apply Lemma 2.23 $m-1-r$ times, i.e.,

$$\begin{aligned}
&F_k(c)A\begin{pmatrix} r,\ldots,m \\ r,\ldots,m \end{pmatrix} - \frac{1}{c}F_{k-1}(c)a_{r,r}A\begin{pmatrix} r+1,\ldots,m \\ r+1,\ldots,m \end{pmatrix} \\
&> a_{rr}\left[F_{k+1}(c)A\begin{pmatrix} r+1,\ldots,m \\ r+1,\ldots,m \end{pmatrix} - \frac{1}{c}F_k(c)a_{r+1,r+1}A\begin{pmatrix} r+2,\ldots,m \\ r+2,\ldots,m \end{pmatrix}\right] \\
&> a_{rr}a_{r+1,r+1}\left[F_{k+2}(c)A\begin{pmatrix} r+2,\ldots,m \\ r+2,\ldots,m \end{pmatrix}\right. \\
&\qquad\qquad \left. -\frac{1}{c}F_{k+1}(c)a_{r+2,r+2}A\begin{pmatrix} r+3,\ldots,m \\ r+3,\ldots,m \end{pmatrix}\right] \\
&> \quad \cdots \\
&> a_{rr}a_{r+1,r+1}\cdots a_{m-2,m-2} \\
&\quad \times\left[F_{k+m-1-r}(c)A\begin{pmatrix} m-1,m \\ m-1,m \end{pmatrix} - \frac{1}{c}F_{k+m-2-r}(c)a_{m-1,m-1}a_{mm}\right].
\end{aligned}$$

Now

$$A\begin{pmatrix} m-1,m \\ m-1,m \end{pmatrix} = a_{m-1,m-1}a_{mm} - a_{m-1,m}a_{m,m-1}$$

and as

$$a_{m-1,m-1}a_{mm} > c\,a_{m-1,m}a_{m,m-1}$$

we have

$$\begin{aligned}
A\begin{pmatrix} m-1,m \\ m-1,m \end{pmatrix} &> a_{m-1,m-1}a_{mm} - \frac{1}{c}a_{m-1,m-1}a_{mm} \\
&= \left(1-\frac{1}{c}\right)a_{m-1,m-1}a_{mm}.
\end{aligned}$$

Substituting in the above gives

$$\begin{aligned} F_k(c)A\begin{pmatrix} r,\dots,m \\ r,\dots,m \end{pmatrix} - \frac{1}{c}F_{k-1}(c)a_{r,r}A\begin{pmatrix} r+1,\dots,m \\ r+1,\dots,m \end{pmatrix} \\ > a_{rr}a_{r+1,r+1}\cdots a_{m-2,m-2} \\ \left[F_{k+m-1-r}(c)\left(1-\frac{1}{c}\right)a_{m-1,m-1}a_{mm}\right. \\ \left. -\frac{1}{c}F_{k+m-2-r}(c)a_{m-1,m-1}a_{mm}\right] \\ = a_{rr}a_{r+1,r+1}\cdots a_{mm}\left[F_{k+m-1-r}(c)\left(1-\frac{1}{c}\right)-\frac{1}{c}F_{k+m-2-r}(c)\right]. \end{aligned}$$

From the recurrence relation of Lemma 2.20,

$$\begin{aligned} F_{k+m-r+1}(c) &= F_{k+m-r}(c) - \frac{1}{c}F_{k+m-r-1}(c) \\ &= F_{k+m-r-1}(c) - \frac{1}{c}F_{k+m-r-2}(c) - \frac{1}{c}F_{k+m-r-1}(c). \end{aligned}$$

Thus

$$\begin{aligned} F_k(c)A\begin{pmatrix} r,\dots,m \\ r,\dots,m \end{pmatrix} - \frac{1}{c}F_{k-1}(c)a_{r,r}A\begin{pmatrix} r+1,\dots,m \\ r+1,\dots,m \end{pmatrix} \\ > a_{rr}a_{r+1,r+1}\cdots a_{mm}F_{k+m-r+1}(c). \end{aligned}$$

□

Lemma 2.25 *Under Assumption A and if $F_k(c) > 0$, $k = 1,\dots,s-1$, then*

$$A\begin{pmatrix} 1,\dots,m \\ 1,\dots,m \end{pmatrix} >$$

$$a_{11}\cdots a_{ss}\left[F_s(c)A\begin{pmatrix} s+1,\dots,m \\ s+1,\dots,m \end{pmatrix} - \frac{1}{c}F_{s-1}(c)a_{s+1,s+1}A\begin{pmatrix} s+2,\dots,m \\ s+2,\dots,m \end{pmatrix}\right]$$

for $s = 1,\dots,m-2$.

Proof We start with $r = 1$ in Lemma 2.22, which can be written as

$$\begin{aligned} A\begin{pmatrix} 1,\dots,m \\ 1,\dots,m \end{pmatrix} &> a_{11}\left[A\begin{pmatrix} 2,\dots,m \\ 2,\dots,m \end{pmatrix} - \frac{1}{c}a_{22}A\begin{pmatrix} 3,\dots,m \\ 3,\dots,m \end{pmatrix}\right] \\ &= a_{11}\left[F_1(c)A\begin{pmatrix} 2,\dots,m \\ 2,\dots,m \end{pmatrix} - \frac{1}{c}F_0(c)a_{22}A\begin{pmatrix} 3,\dots,m \\ 3,\dots,m \end{pmatrix}\right]. \end{aligned}$$

We now apply Lemma 2.23 $s-1$ times for $r = 2,\dots,s$, with $k = r-1$, to obtain our result. □

We are now in a position to prove the sufficiency part of Theorem 2.16. The induction hypothesis will be that if $A = (a_{ij})$ is an $m \times m$ ($m \geq 3$) matrix with strictly positive entries, for which

$$a_{ij}a_{i+1,j+1} > c_m\, a_{i,j+1}a_{i+1,j}, \tag{2.14}$$

for $i, j = 1, \ldots, m-1$, where $c_m = 4\cos^2 \frac{\pi}{m+1}$; then A is strictly totally positive and

$$A\begin{pmatrix}1,\ldots,m\\1,\ldots,m\end{pmatrix} > a_{11}A\begin{pmatrix}2,\ldots,m\\2,\ldots,m\end{pmatrix} - a_{12}a_{21}A\begin{pmatrix}3,\ldots,m\\3,\ldots,m\end{pmatrix}.$$

We first prove that this result holds in the case $m = 3$. It is easily verified that the inequality

$$A\begin{pmatrix}1,2,3\\1,2,3\end{pmatrix} > a_{11}A\begin{pmatrix}2,3\\2,3\end{pmatrix} - a_{12}a_{21}A\begin{pmatrix}3\\3\end{pmatrix}$$

holds for every strictly totally positive matrix. Thus one only need verify (since $c_3 = 2$) that if A is a 3×3 matrix with strictly positive entries, for which

$$a_{ij}a_{i+1,j+1} > 2\, a_{i,j+1}a_{i+1,j},$$

for $i, j = 1, 2$, then A is strictly totally positive. That is, we must verify that $\det A > 0$. Now

$$\begin{aligned}\det A &= a_{11}A\begin{pmatrix}2,3\\2,3\end{pmatrix} - a_{12}A\begin{pmatrix}2,3\\1,3\end{pmatrix} + a_{13}A\begin{pmatrix}2,3\\1,2\end{pmatrix}\\ &> a_{11}A\begin{pmatrix}2,3\\2,3\end{pmatrix} - a_{12}A\begin{pmatrix}2,3\\1,3\end{pmatrix}.\end{aligned}$$

From (2.14) we have

$$a_{11}a_{22} > 2\, a_{12}a_{21}, \qquad a_{22}a_{33} > 2\, a_{23}a_{32}.$$

Thus

$$a_{11}a_{22}a_{33} > a_{12}a_{21}a_{33} + a_{11}a_{23}a_{32},$$

from which

$$\begin{aligned}a_{11}A\begin{pmatrix}2,3\\2,3\end{pmatrix} &= a_{11}a_{22}a_{33} - a_{11}a_{23}a_{32} > a_{12}a_{21}a_{33}\\ &> a_{12}a_{21}a_{33} - a_{12}a_{23}a_{31} = a_{12}A\begin{pmatrix}2,3\\1,3\end{pmatrix}.\end{aligned}$$

This proves the result in the case $m = 3$.

We assume that our induction hypothesis is valid for $m < n$ and prove it for $m = n$. We start with the following.

Proposition 2.26 *Under the above assumptions*

$$a_{1j}A\begin{pmatrix} 2,\ldots,n \\ 1,\ldots,\widehat{j},\ldots,n\end{pmatrix} - a_{1,j+1}A\begin{pmatrix} 2,\ldots,n \\ 1,\ldots,\widehat{j+1},\ldots,n\end{pmatrix} > 0,$$

$j = 2,\ldots,n-1$.

Proof We use induction on the $(n-1)\times(n-1)$ submatrices

$$A\begin{bmatrix} 2,\ldots,n \\ 1,\ldots,\widehat{j},\ldots,n\end{bmatrix}, \qquad A\begin{bmatrix} 2,\ldots,n \\ 1,\ldots,\widehat{j+1},\ldots,n\end{bmatrix}.$$

As $c_{n-1} < c_n$ we can apply the induction hypothesis, and thus both matrices are strictly totally positive and also satisfy (2.13) of Assumption A. We consider separately each of the two factors

$$A\begin{pmatrix} 2,\ldots,n \\ 1,\ldots,\widehat{j},\ldots,n\end{pmatrix}, \qquad A\begin{pmatrix} 2,\ldots,n \\ 1,\ldots,\widehat{j+1},\ldots,n\end{pmatrix}.$$

By the Hadamard Inequality (Theorem 1.21)

$$A\begin{pmatrix} 2,\ldots,n \\ 1,\ldots,\widehat{j+1},\ldots,n\end{pmatrix} < a_{21}\cdots a_{j+1,j}A\begin{pmatrix} j+2,\ldots,n \\ j+2,\ldots,n\end{pmatrix}.$$

Furthermore, from Lemma 2.19,

$$a_{1j}a_{j+1,j+1} > c_n^j a_{1,j+1}, a_{j+1,j}.$$

Thus

$$\begin{aligned} a_{1,j+1}A&\begin{pmatrix} 2,\ldots,n \\ 1,\ldots,\widehat{j+1},\ldots,n\end{pmatrix} \\ &< a_{21}\cdots a_{j+1,j}a_{1,j+1}A\begin{pmatrix} j+2,\ldots,n \\ j+2,\ldots,n\end{pmatrix} \\ &< \frac{1}{c_n^j}a_{1j}a_{21}\cdots a_{j,j-1}a_{j+1,j+1}A\begin{pmatrix} j+2,\ldots,n \\ j+2,\ldots,n\end{pmatrix}. \end{aligned} \tag{2.15}$$

Let us now consider

$$A\begin{pmatrix} 2,\ldots,n \\ 1,\ldots,\widehat{j},\ldots,n\end{pmatrix}.$$

As $c_n = 4\cos^2\theta_n$, $\theta_n = \frac{\pi}{n+1}$, and $F_k(c) > 0$ for $c = 4\cos^2\theta$ where $\theta < \frac{\pi}{k+1}$ (see Lemma 2.21(i)), it follows that

$$F_k(c_n) > 0, \qquad k = 1,\ldots,n-1.$$

Thus we can apply Lemma 2.25 with $s = j-2$ to obtain

$$A\begin{pmatrix} 2,\ldots,n \\ 1,\ldots,\widehat{j},\ldots,n \end{pmatrix} > a_{21}\cdots a_{j-1,j-2}\left[F_{j-2}(c_n)A\begin{pmatrix} j,j+1,\ldots,n \\ j-1,j+1,\ldots,n \end{pmatrix}\right.$$
$$\left.-\frac{1}{c_n}F_{j-3}(c_n)a_{j,j-1}A\begin{pmatrix} j+1,\ldots,n \\ j+1,\ldots,n \end{pmatrix}\right].$$

By the induction hypothesis from (2.13)

$$\begin{aligned} &A\begin{pmatrix} j,j+1,\ldots,n \\ j-1,j+1,\ldots,n \end{pmatrix} \\ &\quad > a_{j,j-1}A\begin{pmatrix} j+1,\ldots,n \\ j+1,\ldots,n \end{pmatrix} - a_{j,j+1}a_{j+1,j-1}A\begin{pmatrix} j+2,\ldots,n \\ j+2,\ldots,n \end{pmatrix}, \end{aligned}$$

and thus

$$\begin{aligned} &A\begin{pmatrix} 2,\ldots,n \\ 1,\ldots,\widehat{j},\ldots,n \end{pmatrix} \\ &> a_{21}\cdots a_{j-1,j-2}\left[a_{j,j-1}F_{j-2}(c_n)A\begin{pmatrix} j+1,\ldots,n \\ j+1,\ldots,n \end{pmatrix}\right. \\ &\qquad -a_{j,j+1}a_{j+1,j-1}F_{j-2}(c_n)A\begin{pmatrix} j+2,\ldots,n \\ j+2,\ldots,n \end{pmatrix} \\ &\qquad \left.-\frac{1}{c_n}F_{j-3}(c_n)\,a_{j,j-1}A\begin{pmatrix} j+1,\ldots,n \\ j+1,\ldots,n \end{pmatrix}\right] \\ &= a_{21}\cdots a_{j-1,j-2}\left[a_{j,j-1}\left[F_{j-2}(c_n)-\frac{1}{c_n}F_{j-3}(c_n)\right]A\begin{pmatrix} j+1,\ldots,n \\ j+1,\ldots,n \end{pmatrix}\right. \\ &\qquad \left.-a_{j,j+1}a_{j+1,j-1}F_{j-2}(c_n)A\begin{pmatrix} j+2,\ldots,n \\ j+2,\ldots,n \end{pmatrix}\right]. \end{aligned}$$

Since

$$F_{j-1}(c) = F_{j-2}(c) - \frac{1}{c}F_{j-3}(c)$$

and

$$a_{j,j-1}a_{j+1,j+1} > c_n^2 a_{j,j+1}a_{j+1,j-1}$$

we have

$$A\begin{pmatrix} 2,\ldots,n \\ 1,\ldots,\widehat{j},\ldots,n \end{pmatrix}$$
$$> a_{21}\cdots a_{j-1,j-2}\left[a_{j,j-1}F_{j-1}(c_n)A\begin{pmatrix} j+1,\ldots,n \\ j+1,\ldots,n \end{pmatrix} - \frac{a_{j,j-1}a_{j+1,j+1}}{c_n^2}F_{j-2}(c_n)A\begin{pmatrix} j+2,\ldots,n \\ j+2,\ldots,n \end{pmatrix}\right]$$
$$= a_{21}\cdots a_{j-1,j-2}a_{j,j-1}\left[F_{j-1}(c_n)A\begin{pmatrix} j+1,\ldots,n \\ j+1,\ldots,n \end{pmatrix} - \frac{a_{j+1,j+1}}{c_n^2}F_{j-2}(c_n)A\begin{pmatrix} j+2,\ldots,n \\ j+2,\ldots,n \end{pmatrix}\right].$$

From Lemma 2.22

$$A\begin{pmatrix} j+1,\ldots,n \\ j+1,\ldots,n \end{pmatrix} > a_{j+1,j+1}\left[A\begin{pmatrix} j+2,\ldots,n \\ j+2,\ldots,n \end{pmatrix} - \frac{1}{c_n}\,a_{j+2,j+2}A\begin{pmatrix} j+3,\ldots,n \\ j+3,\ldots,n \end{pmatrix}\right].$$

Thus, finally,

$$A\begin{pmatrix} 2,\ldots,n \\ 1,\ldots,\widehat{j},\ldots,n \end{pmatrix}$$
$$> \; a_{21}\cdots a_{j-1,j-2}a_{j,j-1}a_{j+1,j+1}\left[F_{j-1}(c_n) \left(A\begin{pmatrix} j+2,\ldots,n \\ j+2,\ldots,n \end{pmatrix} - \frac{1}{c_n}a_{j+2,j+2}A\begin{pmatrix} j+3,\ldots,n \\ j+3,\ldots,n \end{pmatrix}\right) - \frac{1}{c_n^2}F_{j-2}(c_n)A\begin{pmatrix} j+2,\ldots,n \\ j+2,\ldots,n \end{pmatrix}\right]$$
$$= \; a_{21}\cdots a_{j-1,j-2}a_{j,j-1}a_{j+1,j+1}\left[\left(F_{j-1}(c_n) - \frac{1}{c_n^2}F_{j-2}(c_n)\right) A\begin{pmatrix} j+2,\ldots,n \\ j+2,\ldots,n \end{pmatrix} - \frac{1}{c_n}a_{j+2,j+2}F_{j-1}(c_n)A\begin{pmatrix} j+3,\ldots,n \\ j+3,\ldots,n \end{pmatrix}\right]. \quad (2.16)$$

We now combine (2.15) and (2.16) to obtain

$$
\begin{aligned}
a_{1j} A\begin{pmatrix} 2,\ldots,n \\ 1,\ldots,\hat{j},\ldots,n \end{pmatrix} &- a_{1,j+1} A\begin{pmatrix} 2,\ldots,n \\ 1,\ldots,\widehat{j+1},\ldots,n \end{pmatrix} \\
&> a_{1j}a_{21}\cdots a_{j-1,j-2}a_{j,j-1}a_{j+1,j+1}\left[\left(F_{j-1}(c_n) - \frac{1}{c_n^2}F_{j-2}(c_n)\right)\right. \\
&\qquad \left. A\begin{pmatrix} j+2,\ldots,n \\ j+2,\ldots,n \end{pmatrix} - \frac{1}{c_n}a_{j+2,j+2}F_{j-1}(c_n)A\begin{pmatrix} j+3,\ldots,n \\ j+3,\ldots,n \end{pmatrix}\right] \\
&\qquad\quad - \frac{1}{c_n^j}a_{1j}a_{21}\cdots a_{j,j-1}a_{j+1,j+1}A\begin{pmatrix} j+2,\ldots,n \\ j+2,\ldots,n \end{pmatrix} \\
&= a_{1j}a_{21}\cdots a_{j-1,j-2}a_{j,j-1}a_{j+1,j+1}\left[\left(F_{j-1}(c_n) - \frac{1}{c_n^2}F_{j-2}(c_n) - \frac{1}{c_n^j}\right)\right. \\
&\qquad \left. A\begin{pmatrix} j+2,\ldots,n \\ j+2,\ldots,n \end{pmatrix} - \frac{1}{c_n}a_{j+2,j+2}F_{j-1}(c_n)A\begin{pmatrix} j+3,\ldots,n \\ j+3,\ldots,n \end{pmatrix}\right].
\end{aligned}
$$

From Lemma 2.21(ii)

$$
F_{j-1}(c_n) - \frac{1}{c_n^2}F_{j-2}(c_n) - \frac{1}{c_n^j} \geq F_j(c_n)
$$

for $n \geq 3$, $j = 2, 3, \ldots, n$. Since for these values of j and n we have $F_j(c_n) \geq 0$, it follows that

$$
\begin{aligned}
a_{1j} A\begin{pmatrix} 2,\ldots,n \\ 1,\ldots,\hat{j},\ldots,n \end{pmatrix} &- a_{1,j+1} A\begin{pmatrix} 2,\ldots,n \\ 1,\ldots,\widehat{j+1},\ldots,n \end{pmatrix} \\
&> a_{1j}a_{21}\cdots a_{j-1,j-2}a_{j,j-1}a_{j+1,j+1}\left[F_j(c_n)A\begin{pmatrix} j+2,\ldots,n \\ j+2,\ldots,n \end{pmatrix}\right. \\
&\qquad \left. - \frac{1}{c_n}a_{j+2,j+2}F_{j-1}(c_n)A\begin{pmatrix} j+3,\ldots,n \\ j+3,\ldots,n \end{pmatrix}\right].
\end{aligned}
$$

From Lemma 2.24 (with $k = j$ and $r = j+2$) the above quantity satisfies

$$
> a_{1j}a_{21}\cdots a_{j-1,j-2}a_{j,j-1}a_{j+1,j+1}a_{j+2,j+2}\cdots a_{nn}F_{n-1}(c_n).
$$

As $F_{n-1}(c_n) > 0$ this proves Proposition 2.26. □

It now remains to advance the induction hypothesis, i.e., prove that A is strictly totally positive and (2.13) holds. This demands proving

$$
A\begin{pmatrix} 1,\ldots,n \\ 1,\ldots,n \end{pmatrix} > 0 \tag{2.17}
$$

and

$$A\binom{1,\ldots,n}{1,\ldots,n} > a_{11}A\binom{2,\ldots,n}{2,\ldots,n} - a_{12}a_{21}A\binom{3,\ldots,n}{3,\ldots,n}. \qquad (2.18)$$

From Proposition 2.26 (which assumed the validity of the result for $(n-1)\times(n-1)$ matrices)

$$\begin{aligned} A\binom{1,\ldots,n}{1,\ldots,n} &= \sum_{k=1}^{n}(-1)^{k+1}a_{1k}A\binom{2,\ldots,n}{1,\ldots,\widehat{k},\ldots,n} \\ &> a_{11}A\binom{2,\ldots,n}{2,\ldots,n} - a_{12}A\binom{2,3,\ldots,n}{1,3,\ldots,n}. \end{aligned}$$

By the Hadamard Inequality

$$A\binom{2,\ldots,n}{1,3,\ldots,n} < a_{21}A\binom{3,\ldots,n}{3,\ldots,n}$$

and thus

$$A\binom{1,\ldots,n}{1,\ldots,n} > a_{11}A\binom{2,\ldots,n}{2,\ldots,n} - a_{12}a_{21}A\binom{3,\ldots,n}{3,\ldots,n}.$$

This proves (2.18).

As (2.12) holds for $r=2,\ldots,n$, by the induction hypothesis, and since we have now proven (2.18), which completes (2.13) for $m=n$ (again applying the induction hypothesis), we have that the results of Lemmas 2.22–2.25 hold for $m=n$. From Lemma 2.25 (with $m=n$ and $s=n-2$)

$$\begin{aligned} & A\binom{1,\ldots,n}{1,\ldots,n} \\ & > a_{11}\cdots a_{n-2,n-2}\left[F_{n-2}(c_n)A\binom{n-1,n}{n-1,n} - \frac{1}{c_n}F_{n-3}(c_n)a_{n-1,n-1}a_{nn}\right]. \end{aligned}$$

Now (see the proof of Lemma 2.24)

$$A\binom{n-1,n}{n-1,n} = a_{n-1,n-1}a_{nn} - a_{n-1,n}a_{n,n-1} > \left(1-\frac{1}{c_n}\right)a_{n-1,n-1}a_{nn}.$$

Thus

$$\begin{aligned} & F_{n-2}(c_n)A\binom{n-1,n}{n-1,n} - \frac{1}{c_n}F_{n-3}(c_n)a_{n-1,n-1}a_{nn} \\ & > F_{n-2}(c_n)\left(1-\frac{1}{c_n}\right)a_{n-1,n-1}a_{nn} - \frac{1}{c_n}F_{n-3}(c_n)a_{n-1,n-1}a_{nn} \\ & = a_{n-1,n-1}a_{nn}\left[F_{n-2}(c_n)\left(1-\frac{1}{c_n}\right) - F_{n-3}(c_n)\frac{1}{c_n}\right]. \end{aligned}$$

From the recurrence relation of Lemma 2.20(ii)

$$\begin{aligned} 0 = F_n(c_n) &= F_{n-1}(c_n) - \frac{1}{c_n}F_{n-2}(c_n) \\ &= F_{n-2}(c_n) - \frac{1}{c_n}F_{n-3}(c_n) - \frac{1}{c_n}F_{n-2}(c_n). \end{aligned}$$

Substituting it follows that

$$A\begin{pmatrix} 1,\ldots,n \\ 1,\ldots,n \end{pmatrix} > 0.$$

This advances the induction and completes the proof of Theorem 2.16.

2.7 Remarks

Fekete's Lemma 2.1 can be found in Fekete, Pólya [1912]. (Different sections of this paper were written by the different authors.) The fact that for strictly totally positive matrices it suffices to consider only minors with consecutive rows and consecutive columns implies that the remaining minors are nonnegative combinations of the minors with consecutive rows and consecutive columns. If we detail the proof of Fekete's Lemma 2.1 we see that the coefficients in this combination depend upon other minors of the same size, plus lower order minors, and in a nonlinear fashion. Such a formula is not valid for totally positive matrices in general. Theorem 2.3 was proved in Gasca, Peña [1992]. Their proof uses factorization, a subject to be considered in Chapter 6 (see also Metelmann [1973]). The two proofs given here are different. Proposition 2.5 first appears in Gasca, Peña [1992]. It is reproved in Gladwell [1998]. Theorem 2.6 is due to Whitney [1952]. Proposition 1.9 is also to be found in Whitney [1952]. Theorem 2.8 appears in Karlin [1968], p. 85. Proposition 2.9 can be found in Shapiro, Shapiro [1995]. Theorem 2.10 is in Cryer [1973]. Theorem 2.12 was conjectured in Cryer [1973] and proved in Cryer [1976] by very different methods. The proof given here can be found in Micchelli, Pinkus [1991]. The example of a 4×4 matrix, which is not totally positive but whose minors with consecutive rows and consecutive columns are nonnegative, is from Cryer [1973]. Theorem 2.13 was proved in Cryer [1976]. Theorem 2.14 was conjectured in Cryer [1973] and proved in Cryer [1976]. Proposition 2.15 is from Gasca, Peña [1993].

In Proposition 1.20, as a consequence of Theorem 1.19, we proved that a nonsingular totally positive matrix A is almost strictly totally positive if

$$A\begin{pmatrix} i+1,\ldots,i+p \\ j+1,\ldots,j+p \end{pmatrix} > 0$$

whenever $a_{i+k,j+k} > 0$, $k = 1, \dots, p$, for all possible i, j and p. As a further consequence of Theorem 1.19, paralleling Proposition 2.5 and Proposition 2.9, it is not difficult to show that for a nonsingular totally positive matrix it is hardly necessary to verify the above conditions for all i, j, and p in order to determine if the matrix A is almost strictly totally positive. If we know the zero entries of A then, from Theorem 1.19, we can determine a minimal number of such conditions that must be verified. This is done using other methods in Gasca, Peña [1995], Gladwell [2004], and Gasca, Peña [2006].

Craven, Csordas [1998] first proved that if $A = (a_{ij})$ is an $n \times m$ matrix with strictly positive entries and

$$a_{ij}a_{i+1,j+1} \geq c^* \, a_{i,j+1}a_{i+1,j}$$

for all $i = 1, \dots, n-1$, $j = 1, \dots, m-1$, where $c^* = 4.07\dots$ is the unique real root of $x^3 - 5x^2 + 4x - 1 = 0$, then A is strictly totally positive. The fact that there is such a universal constant was in itself a surprising result. There was no claim made in the paper that their c^* was best possible. On the other hand it is not difficult to show that we must have $c^* \geq 4$. Dimitrov, Peña [2005] conjectured that $c^* = 4$, in general, and $c_n = 4\cos^2 \frac{\pi}{n+1}$ is best possible for each $n \times n$ matrix. Unaware of Dimitrov, Peña [2005], this conjecture was proved by Katkova, Vishnyakova [2006]. It is their proof that is presented here. A simpler proof would be of interest.

3
Variation diminishing

The variation diminishing property of a matrix was introduced by I. J. Schoenberg in 1930. In Section 3.1 we prove two fundamental results concerning variation diminishing properties of totally positive matrices. The term "variation diminishing" was coined by Pólya to describe the property whereby the number of sign changes of $A\mathbf{x}$ is bounded above by the number of sign changes of $\mathbf{x}$. The term is somewhat of a misnomer since it is not the "variation" that does not increase. The variation diminishing property and the exact form in which these bounds hold are the content of the two theorems of Section 3.1. In Section 3.2 we present one application. Other applications can be found in subsequent chapters.

3.1 Main equivalence theorems

To explain the concept of variation diminishing we first fix some notation. We define two counts for the number of sign changes of a vector $\mathbf{c} = (c_1, \ldots, c_n) \in \mathbb{R}^n$. These are

$S^-(\mathbf{c})$ – the number of sign changes in the sequence $c_1, \ldots, c_n$, with zero terms discarded. Obviously $S^-(\mathbf{0}) = 0$.

$S^+(\mathbf{c})$ – the maximum number of sign changes in the sequence $c_1, \ldots, c_n$, where zero terms are arbitrarily assigned values $+1$ or -1. For convenience we set $S^+(\mathbf{0}) = n$.

Thus, for example,

$$S^-(1,0,2,-3,0,1) = 2, \text{ and } S^+(1,0,2,-3,0,1) = 4\,.$$

In our "S^- count" we have a sign change from 2 to -3 and another from -3 to 1. In our "S^+ count," by giving the first (left-most) 0 the value -1, we get two additional sign changes. Giving the second 0 the value 1 or -1 does not alter the number of sign changes.

The following elementary facts will be used. Their proof is left to the reader.

Lemma 3.1 *For every* $\mathbf{c} = (c_1, c_2, \ldots, c_n) \in \mathbb{R}^n \backslash \{\mathbf{0}\}$ *we have*

$$S^+(\mathbf{c}) + S^-(\widetilde{\mathbf{c}}) = n - 1,$$

where $\widetilde{\mathbf{c}} = (c_1, -c_2, \ldots, (-1)^{n-1} c_n)$*, i.e., we replace* c_r *in* $\mathbf{c}$ *by* $(-1)^{r-1} c_r$ *for each* r*.*

Lemma 3.2 *If*

$$\lim_{k \to \infty} \mathbf{c}_k = \mathbf{c}$$

then

$$\varliminf_{k \to \infty} S^-(\mathbf{c}_k) \geq S^-(\mathbf{c})$$

and

$$\varlimsup_{k \to \infty} S^+(\mathbf{c}_k) \leq S^+(\mathbf{c}).$$

The first main result is the following.

Theorem 3.3 *Let* A *be an* $n \times m$ *strictly totally positive matrix. Then*

(a) *for each vector* $\mathbf{x} \in \mathbb{R}^m$, $\mathbf{x} \neq \mathbf{0}$,

$$S^+(A\mathbf{x}) \leq S^-(\mathbf{x});$$

(b) *if* $S^+(A\mathbf{x}) = S^-(\mathbf{x})$ *and* $A\mathbf{x} \neq \mathbf{0}$*, then the sign of the first (and last) component of* $A\mathbf{x}$ *(if zero, the sign given in determining* $S^+(A\mathbf{x})$*) agrees with the sign of the first (and last) nonzero component of* $\mathbf{x}$.

Conversely, if (a) *and* (b) *hold for some* $n \times m$ *matrix* A *and every* $\mathbf{x} \in \mathbb{R}^m$, $\mathbf{x} \neq \mathbf{0}$*, then* A *is strictly totally positive.*

Proof We first prove that if A is strictly totally positive then (a) and (b) hold.

Assume $\mathbf{x} \neq \mathbf{0}$ and $S^-(\mathbf{x}) = r$. The components of $\mathbf{x}$ may be therefore divided into $r + 1$ blocks

$$(x_1, \ldots, x_{s_1}), (x_{s_1+1}, \ldots, x_{s_2}), \ldots, (x_{s_r+1}, \ldots, x_m)$$

where, without loss of generality, we assume that

$$x_k(-1)^{i+1} \geq 0, \qquad k = s_{i-1}+1, \ldots, s_i,$$

for $i = 1, \ldots, r+1$, for each i strict inequality holds for at least one k, and where we set $s_0 = 0$ and $s_{r+1} = m$. Let $\mathbf{a}^1, \ldots, \mathbf{a}^m \in \mathbb{R}^n$ denote the column vectors of A and set

$$\mathbf{y}^i = \sum_{k=s_{i-1}+1}^{s_i} |x_k| \mathbf{a}^k, \qquad i = 1, \ldots, r+1.$$

Thus

$$A\mathbf{x} = \sum_{k=1}^{m} x_k \mathbf{a}^k = \sum_{i=1}^{r+1} (-1)^{i+1} \mathbf{y}^i.$$

Let Y be the $n \times (r+1)$ matrix with columns $\mathbf{y}^1, \ldots, \mathbf{y}^{r+1}$. Set

$$\mathbf{w} = A\mathbf{x} = Y\mathbf{d},$$

where $\mathbf{d} = (1, -1, \ldots, (-1)^r) \in \mathbb{R}^{r+1}$. Now a simple calculation shows that

$$Y\begin{pmatrix} i_1, \ldots, i_p \\ j_1, \ldots, j_p \end{pmatrix} = \sum_{k_1=s_{j_1-1}+1}^{s_{j_1}} \cdots \sum_{k_p=s_{j_p-1}+1}^{s_{j_p}} |x_{k_1}| \cdots |x_{k_p}| A\begin{pmatrix} i_1, \ldots, i_p \\ k_1, \ldots, k_p \end{pmatrix}.$$

As

$$A\begin{pmatrix} i_1, \ldots, i_p \\ k_1, \ldots, k_p \end{pmatrix} > 0$$

for all choices of $1 \leq i_1 < \cdots < i_p \leq n$, $1 \leq k_1 < \cdots < k_p \leq m$, and

$$|x_{k_1}| \cdots |x_{k_p}| > 0$$

for some choice of admissible $\{k_1, \ldots, k_p\}$ in the above sum, it follows that Y is an $n \times (r+1)$ strictly totally positive matrix.

We first prove that $S^+(A\mathbf{x}) \leq S^-(\mathbf{x})$. If $n \leq r+1$ there is nothing to prove. For if $A\mathbf{x} \neq \mathbf{0}$, then

$$S^+(A\mathbf{x}) \leq n-1 \leq r = S^-(\mathbf{x}).$$

If $n = r+1$ then $A\mathbf{x} \neq \mathbf{0}$ since $A\mathbf{x} = Y\mathbf{d}$ where Y is a nonsingular $(r+1) \times (r+1)$ matrix and $\mathbf{d} \neq \mathbf{0}$. Thus, if $A\mathbf{x} = \mathbf{0}$ then $n \leq r$ and

$$S^+(A\mathbf{x}) = n \leq r = S^-(\mathbf{x}).$$

We shall therefore assume that $n > r+1$. If $S^+(A\mathbf{x}) \geq r+1$, then there exist indices $1 \leq i_1 < \cdots < i_{r+2} \leq n$ such that

$$\varepsilon w_{i_j}(-1)^{j+1} \geq 0, \qquad j = 1, \ldots, r+2,$$

for some $\varepsilon \in \{-1,1\}$ where $\mathbf{w} = A\mathbf{x} = Y\mathbf{d}$. Since Y is of rank $r+1$, at least two of the w_{i_j} are not zero. Consider the determinant

$$\begin{vmatrix} w_{i_1} & y_{i_1 1} & \cdots & y_{i_1 r+1} \\ \vdots & \vdots & & \vdots \\ w_{i_{r+2}} & y_{i_{r+2} 1} & \cdots & y_{i_{r+2} r+1} \end{vmatrix}.$$

This determinant is zero since the first column is a linear combination of the other columns. Moreover expanding by this first column gives

$$0 = \sum_{j=1}^{r+2} (-1)^{j+1} w_{i_j} Y \begin{pmatrix} i_1, \ldots, \widehat{i_j}, \ldots, i_{r+2} \\ 1, \ldots, r+1 \end{pmatrix}.$$

This equality cannot possibly hold since Y is strictly totally positive and $w_{i_j}(-1)^{j+1}$ is of one fixed (weak) sign and some of the w_{i_j} are not zero. This contradiction implies that $S^+(A\mathbf{x}) \le r = S^-(\mathbf{x})$.

It remains to prove (b), namely that if $S^+(A\mathbf{x}) = S^-(\mathbf{x})$ and $A\mathbf{x} \ne \mathbf{0}$, then the sign of the first (and last) component of $A\mathbf{x}$ (if zero, the sign given in determining $S^+(A\mathbf{x})$) agrees with the sign of the first (and last) nonzero component of $\mathbf{x}$. We continue our analysis using the above notation. We wish to prove that if $S^+(A\mathbf{x}) = S^-(\mathbf{x}) = r$, $A\mathbf{x} \ne \mathbf{0}$, and

$$\varepsilon w_{i_j}(-1)^{j+1} \ge 0, \qquad j = 1, \ldots, r+1,$$

where $\mathbf{w} = Y\mathbf{d} = A\mathbf{x}$, then $\varepsilon = 1$. Since every set of $r+1$ rows of Y is linearly independent, we can solve for any component of $\mathbf{d}$ based on the values $w_{i_1}, \ldots, w_{i_{r+1}}$. Thus

$$1 = \frac{\begin{vmatrix} w_{i_1} & y_{i_1 2} & \cdots & y_{i_1 r+1} \\ \vdots & \vdots & & \vdots \\ w_{i_{r+1}} & y_{i_{r+1} 2} & \cdots & y_{i_{r+1} r+1} \end{vmatrix}}{Y \begin{pmatrix} i_1, \ldots, i_{r+1} \\ 1, \ldots, r+1 \end{pmatrix}}.$$

Expanding by the first column gives

$$1 = \frac{\sum_{j=1}^{r+1} w_{i_j}(-1)^{j+1} Y \begin{pmatrix} i_1, \ldots, \widehat{i_j}, \ldots, i_{r+1} \\ 2, \ldots, r+1 \end{pmatrix}}{Y \begin{pmatrix} i_1, \ldots, i_{r+1} \\ 1, \ldots, r+1 \end{pmatrix}}.$$

As

$$\varepsilon w_{i_j}(-1)^{j+1} \ge 0, \qquad j = 1, \ldots, r+1,$$

and Y is strictly totally positive, we obtain $\varepsilon = 1$.

We now prove the converse direction. Its proof is surprisingly simple. We assume that $S^+(A\mathbf{x}) \le S^-(\mathbf{x})$ for every $\mathbf{x} \ne \mathbf{0}$, and if $S^+(A\mathbf{x}) = S^-(\mathbf{x})$ and $A\mathbf{x} \ne \mathbf{0}$, then the sign of the first (and last) component of $A\mathbf{x}$ (if zero, the sign given in determining $S^+(A\mathbf{x})$) agrees with the sign of the first (and last) nonzero component of $\mathbf{x}$. We shall prove that A is a strictly totally positive matrix.

Our proof is via induction on the size of the minors. We start as follows. Take $\mathbf{e}^j$, the jth unit vector in $\mathbb{R}^m$. As $S^-(\mathbf{e}^j) = 0$ and its nonzero component is positive, we must have that $A\mathbf{e}^j$ is a strictly positive vector. Thus the elements of A are all strictly positive.

Now, assume that all $(r-1) \times (r-1)$ minors of A are strictly positive, $2 \le r \le \min\{n, m\}$, and consider the $r \times r$ submatrix

$$B = A\begin{bmatrix} i_1, \ldots, i_r \\ j_1, \ldots, j_r \end{bmatrix}.$$

If

$$A\begin{pmatrix} i_1, \ldots, i_r \\ j_1, \ldots, j_r \end{pmatrix} = 0$$

then there exists a $\mathbf{z}' = (z_{j_1}, \ldots, z_{j_r}) \in \mathbb{R}^r \backslash \{\mathbf{0}\}$ such that $B\mathbf{z}' = \mathbf{0}$. Let $\mathbf{z} \in \mathbb{R}^m$ be the vector whose j_k component is the above z_{j_k}, $k = 1, \ldots, r$, and whose other components are zero. Then $S^-(\mathbf{z}) \le r-1$ while $S^+(A\mathbf{z}) \ge r$. This is a contradiction. Thus each $r \times r$ minor of A is nonsingular.

We now let $\mathbf{x}' = (x_{j_1}, \ldots, x_{j_r}) \in \mathbb{R}^r$ satisfy $B\mathbf{x}' = \mathbf{d}$ where

$$\mathbf{d} = (1, -1, \ldots, (-1)^{r+1}).$$

Let $\mathbf{x} \in \mathbb{R}^m$ be the vector whose j_k component is the above x_{j_k}, $k = 1, \ldots, r$, and whose other components are zero. Note that $S^-(\mathbf{x}) \le r-1$. From this construction of $\mathbf{x}$ we have $(A\mathbf{x})_{i_k} = (-1)^{k+1}$, $k = 1, \ldots, r$. Thus $S^+(A\mathbf{x}) \ge r-1$. Since, by assumption, $S^-(A\mathbf{x}) \le S^-(\mathbf{x})$ we necessarily have

$$S^-(A\mathbf{x}) = S^-(\mathbf{x}) = r-1$$

and the sign patterns must also agree. Thus

$$(-1)^{k+1}x_{j_k} > 0, \qquad k = 1, \ldots, r.$$

We now return to the equations $B\mathbf{x}' = \mathbf{d}$ and solve for x_{j_1} to obtain

$$0 < x_{j_1} = \frac{\sum_{\ell=1}^{r} A\begin{pmatrix} i_1, \ldots, \widehat{i_\ell}, \ldots, i_r \\ j_2, \ldots, j_r \end{pmatrix}}{A\begin{pmatrix} i_1, \ldots, i_r \\ j_1, \ldots, j_r \end{pmatrix}}.$$

From the induction hypothesis the numerator is positive. Thus

$$A\begin{pmatrix} i_1, \ldots, i_r \\ j_1, \ldots, j_r \end{pmatrix} > 0. \qquad \square$$

For totally positive matrices the result is slightly weaker, but of the same form.

Theorem 3.4 *Let A be an $n \times m$ totally positive matrix. Then*

(a) *for each vector $\mathbf{x} \in \mathbb{R}^m$,*

$$S^-(A\mathbf{x}) \le S^-(\mathbf{x});$$

(b) *if $S^-(A\mathbf{x}) = S^-(\mathbf{x})$ and $A\mathbf{x} \ne \mathbf{0}$, then the sign of the first (and last) nonzero component of $A\mathbf{x}$ agrees with the sign of the first (and last) nonzero component of $\mathbf{x}$.*

Conversely, if (a) *and* (b) *hold for some $n \times m$ matrix A and every $\mathbf{x} \in \mathbb{R}^m$, then A is totally positive.*

Proof Assume A is an $n \times m$ totally positive matrix. From Theorem 2.6 there exists a sequence of $n \times m$ strictly totally positive matrices (A_k) for which

$$\lim_{k\to\infty} A_k = A.$$

Assume $\mathbf{x} \ne \mathbf{0}$. Then from Theorem 3.3

$$S^+(A_k\mathbf{x}) \le S^-(\mathbf{x}).$$

From Lemma 3.2 it follows that

$$S^-(A\mathbf{x}) \le \varliminf_{k\to\infty} S^-(A_k\mathbf{x}).$$

As $S^-(A_k\mathbf{x}) \le S^+(A_k\mathbf{x})$ always holds this implies

$$S^-(A\mathbf{x}) \le S^-(\mathbf{x}),$$

which is (a).

If $S^-(A\mathbf{x}) = S^-(\mathbf{x}) = r$ and $A\mathbf{x} \neq \mathbf{0}$, then for all k sufficiently large we necessarily have, using Lemma 3.2 and Theorem 3.3,

$$r = S^-(A\mathbf{x}) \leq S^-(A_k\mathbf{x}) \leq S^+(A_k\mathbf{x}) \leq S^-(\mathbf{x}) = r,$$

and thus

$$r = S^-(A\mathbf{x}) = S^-(A_k\mathbf{x}) = S^+(A_k\mathbf{x}) = S^-(\mathbf{x}) = r.$$

Since, for k sufficiently large, $S^-(A_k\mathbf{x}) = S^+(A_k\mathbf{x})$ the sign patterns of $A_k\mathbf{x}$ do not depend upon any zero entries. From Theorem 3.3 the sign patterns of $A_k\mathbf{x}$ agree with those of $\mathbf{x}$. From a limiting argument these nonzero sign patterns of $A\mathbf{x}$ and $A_k\mathbf{x}$ agree. This proves (b).

The proof of the converse direction is essentially the same as the corresponding proof in Theorem 3.3. We assume that $S^-(A\mathbf{x}) \leq S^-(\mathbf{x})$ for every $\mathbf{x}$, and if $S^-(A\mathbf{x}) = S^-(\mathbf{x})$ then the sign of the first (and last) component of $A\mathbf{x}$ agrees with the sign of the first (and last) nonzero component of $\mathbf{x}$. We shall prove that A is a totally positive matrix.

Our proof is by induction on the size of the minors. We start as follows. Take $\mathbf{e}^j$, the jth unit vector in $\mathbb{R}^m$. As $S^-(\mathbf{e}^j) = 0$ and its nonzero component is positive we then have that $A\mathbf{e}^j$ is a nonnegative vector. Thus the elements of A are all nonnegative. Now assume that all $(r-1)\times(r-1)$ minors are nonnegative, $2 \leq r \leq \min\{n, m\}$, and consider the $r \times r$ submatrix

$$B = A\begin{bmatrix} i_1, \ldots, i_r \\ j_1, \ldots, j_r \end{bmatrix}.$$

If B is singular, there is nothing to prove. Assume B is nonsingular. We now repeat the argument of the previous theorem.

Let $\mathbf{x}' = (x_{j_1}, \ldots, x_{j_r}) \in \mathbb{R}^r \backslash \{\mathbf{0}\}$ satisfy $B\mathbf{x}' = \mathbf{d}$ where

$$\mathbf{d} = (1, -1, \ldots, (-1)^{r+1}).$$

Let $\mathbf{x} \in \mathbb{R}^m$ be the vector whose j_k component is the above x_{j_k}, $k = 1, \ldots, r$, and whose other components are zero. Note that $S^-(\mathbf{x}) \leq r-1$. From this construction of $\mathbf{x}$ we have $(A\mathbf{x})_{i_k} = (-1)^{k+1}$, $k = 1, \ldots, r$. Thus $S^-(A\mathbf{x}) \geq r-1$. Since, by assumption, $S^-(A\mathbf{x}) \leq S^-(\mathbf{x})$ we necessarily have

$$S^-(A\mathbf{x}) = S^-(\mathbf{x}) = r - 1$$

and the sign patterns must also agree. Thus

$$(-1)^{k+1}x_{j_k} > 0, \qquad k = 1, \ldots, r.$$

We now return to the equations $B\mathbf{x}' = \mathbf{d}$ and solve for x_{j_1} to obtain

$$0 < x_{j_1} = \frac{\sum_{\ell=1}^{r} A\begin{pmatrix} i_1, \ldots, \widehat{i_\ell}, \ldots, i_r \\ j_2, \ldots, j_r \end{pmatrix}}{A\begin{pmatrix} i_1, \ldots, i_r \\ j_1, \ldots, j_r \end{pmatrix}}.$$

From the induction hypothesis the numerator is positive. Thus

$$A\begin{pmatrix} i_1, \ldots, i_r \\ j_1, \ldots, j_r \end{pmatrix} > 0.$$

□

We also have the following if A is both totally positive and nonsingular.

Corollary 3.5 *Let A be an $n \times n$ nonsingular totally positive matrix. Then for each $\mathbf{x} \in \mathbb{R}^n$*

$$S^+(A\mathbf{x}) \le S^+(\mathbf{x}).$$

Proof Let D denote the $n \times n$ diagonal matrix with diagonal entries alternately 1 and -1. From Proposition 1.6 $DA^{-1}D$ is a nonsingular totally positive matrix. As such

$$S^-(DA^{-1}D\mathbf{y}) \le S^-(\mathbf{y})$$

for all $\mathbf{y} \in \mathbb{R}^n$. In addition, we recall from Lemma 3.1 that

$$S^+(\mathbf{c}) + S^-(D\mathbf{c}) = n - 1$$

for all $\mathbf{c} \in \mathbb{R}^n \backslash \{\mathbf{0}\}$.

We may assume that $\mathbf{x} \neq \mathbf{0}$, as there is nothing to prove if $\mathbf{x} = \mathbf{0}$. Setting $\mathbf{y} = DA\mathbf{x}$ we have $\mathbf{x} = A^{-1}D\mathbf{y}$ and from the above

$$\begin{aligned} S^+(A\mathbf{x}) = S^+(D\mathbf{y}) &= (n-1) - S^-(\mathbf{y}) \le (n-1) - S^-(DA^{-1}D\mathbf{y}) \\ &= S^+(A^{-1}D\mathbf{y}) = S^+(\mathbf{x}). \end{aligned}$$

□

3.2 Intervals of strict total positivity

As shall be seen in subsequent chapters, there are numerous applications of the variation diminishing property. We include here one such application.

Let $B = (b_{ij})$ and $C = (c_{ij})$ be $n \times n$ matrices. We apply an ordering to these matrices. We say that $B \le^* C$ if $(-1)^{i+j}\, b_{ij} \le (-1)^{i+j}\, c_{ij}$ for all i, j. Equivalently, if D is the $n \times n$ diagonal matrix with diagonal entries alternately 1 and -1, then $DBD \leq DCD$ where $\leq$ is the usual entrywise inequality.

Theorem 3.6 *Assume* $B = (b_{ij})$ *and* $C = (c_{ij})$ *are strictly totally positive* $n \times n$ *matrices satisfying*

$$B \leq^* C.$$

If A *is an* $n \times n$ *matrix and*

$$B \leq^* A \leq^* C,$$

then A *is strictly totally positive.*

Proof We prove, under the above assumptions, that A is necessarily nonsingular. By continuity it then follows that the determinant of A is positive. Assume, for the moment, that we have proved this fact and let us see how it implies our theorem.

Recall, from Theorem 2.2 (or Theorem 2.3), that to prove that A is strictly totally positive it suffices to show that

$$A\begin{pmatrix} i+1,\ldots,i+k \\ j+1,\ldots,j+k \end{pmatrix} > 0$$

for $i = 0,\ldots,n-k$, $j = 0,\ldots,n-k$, and $k = 1,\ldots,n$. For each such i,j, and k consider the three $k \times k$ submatrices

$$\begin{aligned} B^{ij} &= B\begin{bmatrix} i+1,\ldots,i+k \\ j+1,\ldots,j+k \end{bmatrix}, \\ A^{ij} &= A\begin{bmatrix} i+1,\ldots,i+k \\ j+1,\ldots,j+k \end{bmatrix}, \\ C^{ij} &= C\begin{bmatrix} i+1,\ldots,i+k \\ j+1,\ldots,j+k \end{bmatrix}. \end{aligned}$$

By assumption

$$B^{ij} \leq^* A^{ij} \leq^* C^{ij}$$

for $i+j$ even, while

$$C^{ij} \leq^* A^{ij} \leq^* B^{ij}$$

for $i+j$ odd. This implies, using an induction assumption, that

$$A\begin{pmatrix} i+1,\ldots,i+k \\ j+1,\ldots,j+k \end{pmatrix} > 0,$$

and the theorem is proved. Thus it suffices to prove the nonsingularity of A.

We therefore assume that A is singular, and all $k \times k$ minors of A are strictly positive for $k = 1, \ldots, n-1$. Let $\mathbf{x} \neq \mathbf{0}$ satisfy $A\mathbf{x} = \mathbf{0}$. As A is of rank $n-1$ and

$$A\begin{pmatrix} 2, \ldots, n \\ 1, \ldots, \widehat{j}, \ldots, n \end{pmatrix} > 0$$

for $j = 1, \ldots, n$, it follows that

$$x_j x_{j+1} < 0, \qquad j = 1, \ldots, n-1.$$

Assume, without loss of generality, that

$$(-1)^j x_j > 0, \qquad j = 1, \ldots, n. \tag{3.1}$$

Since

$$(-1)^{i+j}\, b_{ij} \leq (-1)^{i+j}\, a_{ij}, \qquad i, j = 1, \ldots, n,$$

we have

$$\begin{aligned} (-1)^i \sum_{j=1}^n b_{ij} x_j &= \sum_{j=1}^n (-1)^{i+j} b_{ij} (-1)^j x_j \leq \sum_{j=1}^n (-1)^{i+j} a_{ij} (-1)^j x_j \\ &= (-1)^i \sum_{j=1}^n a_{ij} x_j = 0. \end{aligned}$$

That is,

$$(-1)^i (B\mathbf{x})_i \leq (-1)^i (A\mathbf{x})_i = 0, \qquad i = 1, \ldots, n. \tag{3.2}$$

As B is nonsingular we have $B\mathbf{x} \neq \mathbf{0}$. Thus

$$n - 1 = S^+(B\mathbf{x}) = S^-(\mathbf{x}) = n - 1.$$

However from (3.1) and (3.2) we see that the sign of the first (and last) component of $B\mathbf{x}$ (if zero, the sign given in determining $S^+(B\mathbf{x})$) does not agree with the sign of the first (and last) nonzero component of $\mathbf{x}$. This contradicts Theorem 3.3. Thus A is nonsingular. □

3.3 Remarks

According to Schoenberg [1930] the term *variation diminishing* (variationsvermindernd in German) was coined by Pólya. To obtain the variation diminishing property of a matrix, i.e., $S^+(A\mathbf{x}) \leq S^-(\mathbf{x})$ or $S^-(A\mathbf{x}) \leq S^-(\mathbf{x})$ one does not need that the matrix A be strictly totally positive or totally positive, respectively. If A is an $n \times m$ matrix and $n \geq m$ then A satisfies $S^+(A\mathbf{x}) \leq S^-(\mathbf{x})$ for all $\mathbf{x} \in \mathbb{R}^m \backslash \{\mathbf{0}\}$ if and only if A is strictly sign regular (see e.g., Karlin [1968], p. 219, or Ando [1987], p. 29). An $n \times m$ matrix A is *strictly sign regular* if all its minors of order k are of one fixed sign $\varepsilon_k \in \{-1, 1\}$ for each $k = 1, \ldots, m$. If we permit minors to also vanish, then A is said to be *sign regular*. An $n \times m$ matrix A of rank m satisfies $S^-(A\mathbf{x}) \leq S^-(\mathbf{x})$ for all $\mathbf{x} \in \mathbb{R}^m$ if and only if A is sign regular. This was first proven in Schoenberg [1930] with this rank condition (see also Motzkin [1936] and Schoenberg, Whitney [1951] for the more general case). The inequality $n \geq m$ and the rank condition, or something similar, is truly needed in these results. However, for strictly totally positive and totally positive matrices we have property (b) in the statements of Theorems 3.3 and 3.4. This property obviates the need for the rank condition and the inequality $n \geq m$. This fact seems to be generally overlooked. In this regard see Brown, Johnstone, MacGibbon [1981], Lemma A.1, and Fan [1966], Theorems 5 and 6.

For a generalization of sorts of variation diminishing, see Carnicer, Goodman, Peña [1995].

Garloff [1982] proved Theorem 3.6 by different methods. This theorem cannot hold for arbitrary totally positive matrices B and C. (Let B and C be matrices all of whose entries are 0 except for a single 1, suitably chosen.) In Garloff [1982] it is conjectured that the theorem holds if B and C are totally positive and nonsingular. In Garloff [1982] and Garloff [2003] the conjecture is proved for specific subclasses of such matrices.

4
Examples

In this chapter we present various examples of totally positive matrices. Some are specific examples, e.g., exponentials, powers, the Cauchy matrix, and the binomial coefficients. Others are examples of more general classes of matrices where we present conditions determining total positivity within the class, e.g., Green, Jacobi, Hankel, Toeplitz and Hurwitz matrices. Totally positive matrices are not closed under the operation of Hadamard products. In the last section we consider subclasses of totally positive matrices that are closed under this operation. We start, however, by discussing the connection between totally positive matrices, on the one hand, and totally positive kernels and Descartes systems, on the other hand.

4.1 Totally positive kernels and Descartes systems

There are two types of continuous extensions of totally positive or strictly totally positive matrices or, from the opposite perspective, totally positive or strictly totally positive matrices are discrete counterparts of two continuous forms.

A real-valued kernel $K \in C(X \times Y)$ is said to be a *totally positive or strictly totally positive kernel* if

$$(K(x_i, y_j))_{i=1}^{n}\,{}_{j=1}^{m}$$

is, respectively, a totally positive or strictly totally positive matrix for every choice of $x_1 < \cdots < x_n$ in X, $y_1 < \cdots < y_m$ in Y and all possible n and m. If X and Y are finite sets then this is simply a definition of a totally positive or strictly totally positive matrix. As we shall see in the next two sections $K(x, y) = e^{xy}$ is a strictly totally positive kernel on $\mathbb{R} \times \mathbb{R}$, $K(x, y) = x^y$

is a strictly totally positive kernel on $\mathbb{R}_+ \times \mathbb{R}$ and $K(x,y) = 1/(x+y)$ is a strictly totally positive kernel on $\mathbb{R}_+ \times \mathbb{R}_+$.

The semicontinuous analogue of a strictly totally positive matrix, or semidiscrete analogue of a strictly totally positive kernel is given by a *Descartes system*. To define a Descartes system we start with a *Chebyshev system*. A Chebyshev system (generally abbreviated as a T-system) is a system of functions $\{u_1, \ldots, u_r\}$ defined on a set X for which

$$\det\,(u_i(x_j))_{i,j=1}^r \neq 0$$

for every choice of distinct $x_1, \ldots, x_r$ in X. It is called a T_+-system if

$$\det\,(u_i(x_j))_{i,j=1}^r > 0$$

for every choice of $x_1 < \cdots < x_r$ in X. A system of functions $\{u_1, \ldots, u_r\}$ is said to be a Descartes system if every ordered subset of the $u_1, \ldots, u_r$ is a T_+-system. Equivalently

$$(u_i(x_j))_{i=1}^r{}_{j=1}^m$$

is a strictly totally positive matrix for all $x_1 < \cdots < x_m$ in X.

Descartes systems possess a variation diminishing property that follows easily from Theorem 3.3, often termed a *Generalized Descartes Rule of Signs*. It is that if $\{u_1, \ldots, u_r\}$ is a Descartes system on an ordered set X, then for each nontrivial "polynomial"

$$u(x) = \sum_{i=1}^r a_i u_i(x)$$

the number of zeros of u on X is bounded above by $S^-(a_1, \ldots, a_r)$. As will be seen in the next section $\{x^{c_j}\}_{j=1}^r$ is a Descartes system on $\mathbb{R}_+$ for any $c_1 < \cdots < c_r$.

4.2 Exponentials and powers

Here are some elementary but instructive examples of strictly totally positive matrices. Certain specific cases have already been considered (e.g., see the proof of Theorem 2.6).

Let

$$A = (e^{b_i c_j})_{i,j=1}^n$$

where $b_1 < \cdots < b_n$ and $c_1 < \cdots < c_n$. We claim the A is a strictly totally positive matrix. (In other words we are proving that $K(x,y) = e^{xy}$ is a strictly totally positive kernel on $\mathbb{R} \times \mathbb{R}$.) We present a two-step proof. We first prove that the determinant of A is nonzero, and then we prove that it is positive.

Proof If A is singular there exists a $(\lambda_1, \ldots, \lambda_n) \in \mathbb{R}^n \backslash \{\mathbf{0}\}$ for which

$$\sum_{j=1}^{n} \lambda_j e^{b_i c_j} = 0, \qquad i = 1, \ldots, n.$$

Set

$$f(x) = \sum_{j=1}^{n} \lambda_j e^{x c_j}.$$

The function f is a nontrivial exponential polynomial with at least n distinct zeros $b_1, \ldots, b_n$. It is not identically zero since, for example, if m is the largest index for which $\lambda_m \neq 0$, then

$$\lim_{x \to \infty} f(x) e^{-x c_m} = \lambda_m \neq 0.$$

We claim that no such function exists, i.e., every exponential polynomial of this form has at most $n-1$ distinct zeros, hence a contradiction, and A is nonsingular. We prove this claim by induction on n. It obviously holds for $n = 1$. Assume, as above, that f has at least n distinct zeros, $n \geq 2$. Multiply f by $e^{-x c_n}$ and differentiate to obtain

$$g(x) = (f(x) e^{-x c_n})' = \sum_{j=1}^{n-1} \lambda_j (c_j - c_n) e^{x(c_j - c_n)}.$$

The function g is an exponential polynomial with, by Rolle's Theorem, at least $n-1$ zeros. Thus, by our induction hypothesis, g is identically zero, i.e., $\lambda_j = 0$, $j = 1, \ldots, n-1$. Therefore $f(x) = \lambda_n e^{x c_n}$. As f has a zero we have $\lambda_n = 0$. Thus A is nonsingular.

The $(b_i)_{i=1}^n$ and $(c_j)_{j=1}^n$ are arbitrary increasing sequences. As such, proving A nonsingular also proves the nonsingularity of all minors of A. Moreover, by continuity, the sign of the determinant of A is independent of the choice of the arbitrary increasing sequences $(b_i)_{i=1}^n$ and $(c_j)_{j=1}^n$, but may depend upon n. We prove that it is positive for each and every n.

Here are two proofs of the positivity of the determinant of A.

Assume that all $k \times k$ determinants of the form of A are strictly positive for $k = 1, \ldots, n-1$. (The case $k = 1$ is obvious.) Let

$$f(x) = \sum_{j=1}^{n} \lambda_j e^{x c_j}$$

satisfy $f(b_i) = 0$, $i = 1, \ldots, n-1$ and $f(b_n) = 1$. Such an f exists and is unique by the nonsingularity of the associated matrix. In addition,

$$\lambda_n = \frac{\det(e^{b_i c_j})_{i,j=1}^{n-1}}{\det(e^{b_i c_j})_{i,j=1}^{n}}.$$

We proved earlier that f has at most $n-1$ zeros. Since $f(b_n) = 1$ it therefore follows that f is strictly positive for all $x > b_{n-1}$. As such

$$\lambda_n = \lim_{x \to \infty} f(x) e^{-xc_n} > 0.$$

By our induction hypothesis

$$\det(e^{b_i c_j})_{i,j=1}^{n-1} > 0$$

and therefore

$$\det(e^{b_i c_j})_{i,j=1}^{n} > 0.$$

A second proof uses the fact that the sign of the determinant of A is independent of the choice of the arbitrary increasing sequences $(b_i)_{i=1}^n$ and $(c_j)_{j=1}^n$, but may depend upon n. As such it suffices to explicitly calculate the determinant of A for some specific choice of the $(b_i)_{i=1}^n$ and $(c_j)_{j=1}^n$. Let $c_j = j-1$ and set $e^{b_i} = x_i$. Then we have

$$A = (x_i^j)_{i=1}^{n}{}_{j=0}^{n-1}$$

where $0 < x_1 < \cdots < x_n$. This is the well-known Vandermonde matrix whose determinant is given by

$$\det A = \prod_{1 \le \ell < k \le n} (x_k - x_\ell) > 0. \qquad \square$$

Note that with the above substitution $e^{b_i} = x_i$ we also have that

$$A = (x_i^{c_j})_{i,j=1}^n$$

is strictly totally positive for any $0 < x_1 < \cdots < x_n$ and $c_1 < \cdots < c_n$.

There is another method of proof of the total positivity of the $(e^{b_i c_j})$ (for $b_i, c_j > 0$) based on the following idea. Let

$$G(x, y) = \sum_{k=0}^{m} \lambda_k x^k y^k,$$

with $\lambda_k \ge 0$, $k = 0, 1, \ldots, m$. (We assume here that m is finite, but that need not be the case, and is not the case for $G(x,y) = e^{xy}$.) Then for any $0 < b_1 < \cdots < b_n$ and $0 < c_1 < \cdots < c_n$ the matrix

$$A = (G(b_i, c_j))_{i,j=1}^n$$

is totally positive. This follows easily from the fact that

$$A = \begin{pmatrix} b_1^0 & \dots & b_1^m \\ \vdots & \ddots & \vdots \\ b_n^0 & \dots & b_n^m \end{pmatrix} \begin{pmatrix} \lambda_0 & \dots & 0 \\ \vdots & \ddots & \vdots \\ 0 & \dots & \lambda_n \end{pmatrix} \begin{pmatrix} c_1^0 & \dots & c_n^0 \\ \vdots & \ddots & \vdots \\ c_1^m & \dots & c_n^m \end{pmatrix}.$$

Set

$$\begin{aligned} B &= \begin{pmatrix} b_1^0 & \dots & b_1^m \\ \vdots & \ddots & \vdots \\ b_n^0 & \dots & b_n^m \end{pmatrix}, \\ \Lambda &= \begin{pmatrix} \lambda_0 & \dots & 0 \\ \vdots & \ddots & \vdots \\ 0 & \dots & \lambda_n \end{pmatrix}, \\ C &= \begin{pmatrix} c_1^0 & \dots & c_n^0 \\ \vdots & \ddots & \vdots \\ c_1^m & \dots & c_n^m \end{pmatrix}. \end{aligned}$$

Thus

$$A = B\Lambda C.$$

As B and C are strictly totally positive (they are Vandermonde matrices) and Λ is a diagonal totally positive matrix, the result follows.

In this way it can be proven that

$$G_1(x,y) = \cosh xy = \frac{e^{xy} + e^{-xy}}{2} = \sum_{k=0}^{\infty} \frac{x^{2k} y^{2k}}{(2k)!}$$

and

$$G_2(x,y) = \sinh xy = \frac{e^{xy} - e^{-xy}}{2} = \sum_{k=0}^{\infty} \frac{x^{2k+1} y^{2k+1}}{(2k+1)!}$$

are strictly totally positive on $\mathbb{R}_+ \times \mathbb{R}_+$. These facts can also be proven directly as a consequence of the possible number of zeros of any linear combination of the exponentials $e^{xc_1}, e^{-xc_1}, \dots, e^{xc_m}, e^{-xc_m}$ and the parity of $\cosh xy$ and $\sinh xy$.

4.3 Cauchy matrix

The matrix

$$A = \left(\frac{1}{x_i + y_j}\right)_{i,j=1}^{n}$$

is called a *Cauchy matrix*. For $0 < x_1 < \cdots < x_n$, $0 < y_1 < \cdots < y_n$ it is strictly totally positive.

We provide two proofs of this result.

Proof I There is an explicit formula for the determinant of the above matrix and it is given by

$$\det \left(\frac{1}{x_i + y_j}\right)_{i,j=1}^{n} = \frac{\prod_{1\le \ell < k \le n}(x_k - x_\ell)\prod_{1\le \ell < k \le n}(y_k - y_\ell)}{\prod_{i,j=1}^{n}(x_i + y_j)}.$$

Here is a proof of this result. Obviously the result holds for $n = 1$. Now

$$\begin{vmatrix} \frac{1}{x_1+y_1} & \frac{1}{x_1+y_2} & \cdots & \frac{1}{x_1+y_n} \\ \frac{1}{x_2+y_1} & \frac{1}{x_2+y_2} & \cdots & \frac{1}{x_2+y_n} \\ \vdots & \vdots & \ddots & \vdots \\ \frac{1}{x_n+y_1} & \frac{1}{x_n+y_2} & \cdots & \frac{1}{x_n+y_n} \end{vmatrix} = \frac{1}{x_1+y_1} \begin{vmatrix} 1 & \frac{x_1+y_1}{x_1+y_2} & \cdots & \frac{x_1+y_1}{x_1+y_n} \\ \frac{1}{x_2+y_1} & \frac{1}{x_2+y_2} & \cdots & \frac{1}{x_2+y_n} \\ \vdots & \vdots & \ddots & \vdots \\ \frac{1}{x_n+y_1} & \frac{1}{x_n+y_2} & \cdots & \frac{1}{x_n+y_n} \end{vmatrix}.$$

Multiply the first row by $\frac{1}{x_i+y_1}$ and subtract from the ith row to obtain

$$= \frac{1}{x_1+y_1} \begin{vmatrix} 1 & \frac{x_1+y_1}{x_1+y_2} & \cdots & \frac{x_1+y_1}{x_1+y_n} \\ 0 & \frac{(x_2-x_1)(y_2-y_1)}{(x_2+y_2)(x_1+y_2)(x_2+y_1)} & \cdots & \frac{(x_2-x_1)(y_n-y_1)}{(x_2+y_n)(x_1+y_n)(x_2+y_1)} \\ \vdots & \vdots & \ddots & \vdots \\ 0 & \frac{(x_n-x_1)(y_2-y_1)}{(x_n+y_2)(x_1+y_2)(x_n+y_1)} & \cdots & \frac{(x_n-x_1)(y_n-y_1)}{(x_n+y_n)(x_1+y_n)(x_n+y_1)} \end{vmatrix}$$

$$= \frac{\prod_{k=1}^{n}(x_k - x_1)(y_k - y_1)}{(x_1+y_1)\prod_{i=2}^{n}(x_i+y_1)\prod_{j=2}^{n}(x_1+y_j)} \begin{vmatrix} 1 & \frac{x_1+y_1}{x_1+y_2} & \cdots & \frac{x_1+y_1}{x_1+y_n} \\ 0 & \frac{1}{x_2+y_2} & \cdots & \frac{1}{x_2+y_n} \\ \vdots & \vdots & \ddots & \vdots \\ 0 & \frac{1}{x_n+y_2} & \cdots & \frac{1}{x_n+y_n} \end{vmatrix}$$

and now apply an induction argument. □

Proof II We first prove that A is nonsingular.

Assume that $\det A = 0$. There then exist $a_1, \ldots, a_n$, not all zero, for which

$$\sum_{j=1}^{n} \frac{a_j}{x + y_j} = 0, \qquad x = x_1, \ldots, x_n.$$

Thus

$$\frac{\sum_{j=1}^n a_j \left(\prod_{\substack{k=1\\k\neq j}}^n (x+y_k)\right)}{\prod_{j=1}^n (x+y_j)} = 0, \qquad x = x_1,\ldots,x_n,$$

and therefore

$$\sum_{j=1}^n a_j \left(\prod_{\substack{k=1\\k\neq j}}^n (x+y_k)\right) = 0, \qquad x = x_1,\ldots,x_n.$$

The above is a polynomial of degree at most $n-1$ that vanishes at n distinct points and thus is identically zero. In addition the

$$\left\{\prod_{\substack{k=1\\k\neq j}}^n (x+y_k)\right\}_{j=1}^n$$

are linearly independent polynomials since at the value $-y_r$ all but the rth polynomial vanishes. Thus $a_1 = \cdots = a_n = 0$, which is a contradiction, and A is nonsingular.

It remains to prove that each of these determinants is positive. By continuity $\det A$ is of one sign for all choices of $0 < x_1 < \cdots < x_n$, $0 < y_1 < \cdots < y_n$ and a fixed n. For $n = 1$ it is positive by inspection. Let us apply an induction argument.

Multiply the first row of

$$\begin{vmatrix} \frac{1}{x_1+y_1} & \frac{1}{x_1+y_2} & \cdots & \frac{1}{x_1+y_n} \\ \frac{1}{x_2+y_1} & \frac{1}{x_2+y_2} & \cdots & \frac{1}{x_2+y_n} \\ \vdots & \vdots & \ddots & \vdots \\ \frac{1}{x_n+y_1} & \frac{1}{x_n+y_2} & \cdots & \frac{1}{x_n+y_n} \end{vmatrix}$$

by $x_1 + y_1$ to obtain

$$\begin{vmatrix} 1 & \frac{x_1+y_1}{x_1+y_2} & \cdots & \frac{x_1+y_1}{x_1+y_n} \\ \frac{1}{x_2+y_1} & \frac{1}{x_2+y_2} & \cdots & \frac{1}{x_2+y_n} \\ \vdots & \vdots & \ddots & \vdots \\ \frac{1}{x_n+y_1} & \frac{1}{x_n+y_2} & \cdots & \frac{1}{x_n+y_n} \end{vmatrix}.$$

Letting $x_1, y_1 \to 0^+$ the above determinant converges to

$$\begin{vmatrix} \frac{1}{x_2+y_2} & \cdots & \frac{1}{x_2+y_n} \\ \vdots & \ddots & \vdots \\ \frac{1}{x_n+y_2} & \cdots & \frac{1}{x_n+y_n} \end{vmatrix}$$

which, by induction, is positive. □

4.4 Green's matrices

We start not with a Green's matrix but with a simpler related matrix.

Proposition 4.1 *Let $b_1, \ldots, b_n$ be n distinct numbers. Set*

$$a_{ij} = b_{\min\{i,j\}}, \qquad i,j = 1, \ldots, n.$$

Then $A = (a_{ij})_{i,j=1}^n$ is a totally positive matrix if and only if

$$0 \le b_1 < \cdots < b_n.$$

Remark Note that we can always assume that the $\{b_i\}$ are distinct. For if $b_j = b_{j+1}$ then the jth and $(j+1)$st rows and columns are identical.

Proof Assume A is a totally positive matrix. Then $b_k \ge 0$, $k = 1, \ldots, n$. Now

$$0 \le A\begin{pmatrix} i, i+1 \\ i, i+1 \end{pmatrix} = \begin{vmatrix} b_i & b_i \\ b_i & b_{i+1} \end{vmatrix} = b_i(b_{i+1} - b_i).$$

Thus if $b_i > 0$, then $b_{i+1} > b_i$ since the $b_1, \ldots, b_n$ are distinct. If $b_i = 0$, then $b_{i+1} > 0 = b_i$. Therefore

$$0 \le b_1 < \cdots < b_n.$$

If $b_1 = 0$ then the first row and the first column of A are identically zero and we may simply disregard them. As such let us assume that

$$0 < b_1 < \cdots < b_n.$$

Consider

$$A\begin{pmatrix} i_1, \ldots, i_r \\ j_1, \ldots, j_r \end{pmatrix}$$

for $1 \le i_1 < \cdots < i_r \le n$, $1 \le j_1 < \cdots < j_r \le n$. We first claim that if

$$i_1, j_1 < i_2, j_2 < \cdots < i_r, j_r$$

does not hold, then

$$A\begin{pmatrix} i_1, \ldots, i_r \\ j_1, \ldots, j_r \end{pmatrix} = 0.$$

To see this, let us assume, without loss of generality, that $i_\ell \geq j_{\ell+1}$ for some $\ell \in \{1, \ldots, r-1\}$. As $j_1 < \cdots < j_{\ell+1} \leq i_\ell < i_{\ell+1} < \cdots < i_r$ we have

$$a_{i_s j_t} = b_{j_t}$$

for $s = \ell, \ell+1, \ldots, r$; $t = 1, \ldots, \ell+1$. This implies that the $(r-\ell+1)\times(\ell+1)$ submatrix

$$A\begin{bmatrix} i_\ell, i_{\ell+1}, \ldots, i_r \\ j_1, j_2, \ldots, j_{\ell+1} \end{bmatrix}$$

is of rank 1. Thus

$$A\begin{bmatrix} i_1, \ldots, i_r \\ j_1, \ldots, j_{\ell+1} \end{bmatrix}$$

is of rank at most ℓ, i.e., the $\ell+1$ columns thereof are linearly dependent, and therefore

$$A\begin{pmatrix} i_1, \ldots, i_r \\ j_1, \ldots, j_r \end{pmatrix} = 0.$$

Let us now assume that

$$i_1, j_1 < i_2, j_2 < \cdots < i_r, j_r.$$

Set $\alpha_k = \min\{i_k, j_k\}$ and $\beta_k = \max\{i_k, j_k\}$, $k = 1, \ldots, r$. By assumption we have $\alpha_{k+1} > \beta_k$. We will prove that

$$A\begin{pmatrix} i_1, \ldots, i_r \\ j_1, \ldots, j_r \end{pmatrix} = b_{\alpha_1} \prod_{k=1}^{r-1} (b_{\alpha_{k+1}} - b_{\beta_k}) > 0.$$

To this end, note the pattern of this minor, namely

$$A\begin{pmatrix} i_1, \ldots, i_r \\ j_1, \ldots, j_r \end{pmatrix} = \begin{vmatrix} b_{\alpha_1} & b_{i_1} & b_{i_1} & \cdots & b_{i_1} \\ b_{j_1} & b_{\alpha_2} & b_{i_2} & \cdots & b_{i_2} \\ b_{j_1} & b_{j_2} & b_{\alpha_3} & \cdots & b_{i_3} \\ \vdots & \vdots & \vdots & \ddots & \vdots \\ b_{j_1} & b_{j_2} & b_{j_3} & \cdots & b_{\alpha_r} \end{vmatrix}.$$

Multiply the first row by b_{j_1}/b_{α_1} and subtract it from each of the other rows to obtain

$$= \begin{vmatrix} b_{\alpha_1} & b_{i_1} & b_{i_1} & \cdots & b_{i_1} \\ 0 & c_{\alpha_2} & c_{i_2} & \cdots & c_{i_2} \\ 0 & c_{j_2} & c_{\alpha_3} & \cdots & c_{i_3} \\ \vdots & \vdots & \vdots & \ddots & \vdots \\ 0 & c_{j_2} & c_{j_3} & \cdots & c_{\alpha_r} \end{vmatrix}$$

where

$$c_k = b_k - \frac{b_{i_1} b_{j_1}}{b_{\alpha_1}}.$$

For $k > i_1, j_1$, the $\{c_k\}$ is a strictly increasing sequence of positive numbers and we can apply an induction hypothesis (the case $r = 1$ is trivial). Thus

$$A\begin{pmatrix} i_1, \ldots, i_r \\ j_1, \ldots, j_r \end{pmatrix} = b_{\alpha_1} c_{\alpha_2} \prod_{k=2}^{r-1} \left(c_{\alpha_{k+1}} - c_{\beta_k} \right).$$

As

$$c_k = b_k - \frac{b_{i_1} b_{j_1}}{b_{\alpha_1}} = b_k - \frac{b_{\alpha_1} b_{\beta_1}}{b_{\alpha_1}} = b_k - b_{\beta_1},$$

it therefore follows that

$$A\begin{pmatrix} i_1, \ldots, i_r \\ j_1, \ldots, j_r \end{pmatrix} = b_{\alpha_1} \left(b_{\alpha_2} - b_{\beta_1} \right) \prod_{k=2}^{r-1} \left(b_{\alpha_{k+1}} - b_{\beta_k} \right) = b_{\alpha_1} \prod_{k=1}^{r-1} \left(b_{\alpha_{k+1}} - b_{\beta_k} \right).$$

□

Let us now assume that we are given two sequences $c_1, \ldots, c_n$ and $d_1, \ldots, d_n$ of nonzero values. Set

$$a_{ij} = c_{\min\{i,j\}} d_{\max\{i,j\}}, \qquad i, j = 1, \ldots, n.$$

The matrix $A = (a_{ij})_{i,j=1}^n$ is called a *Green's matrix*. We prove the following result.

Theorem 4.2 *The matrix A, as defined above, is totally positive if and only if the $c_1, \ldots, c_n$ and $d_1, \ldots, d_n$ are all of one strict sign and*

$$0 < \frac{c_1}{d_1} \le \cdots \le \frac{c_n}{d_n}.$$

Proof Assume A is totally positive, $n \ge 2$. Then from the positivity of the elements of A we must have that the $c_1, \ldots, c_n$ and $d_1, \ldots, d_n$ are either all positive or all negative. (We are assuming they are all nonzero.) Let us assume, without loss of generality, that they are all positive. Consider

$$0 \le A\begin{pmatrix} i, i+1 \\ i, i+1 \end{pmatrix} = \begin{vmatrix} c_i d_i & c_i d_{i+1} \\ c_i d_{i+1} & c_{i+1} d_{i+1} \end{vmatrix} = c_i d_i d_{i+1}^2 \left(\frac{c_{i+1}}{d_{i+1}} - \frac{c_i}{d_i} \right).$$

Since the $c_1, \ldots, c_n$ and $d_1, \ldots, d_n$ are all positive we have

$$\frac{c_i}{d_i} \le \frac{c_{i+1}}{d_{i+1}}, \qquad i = 1, \ldots, n-1.$$

To prove the converse direction note that since

$$\frac{c_i}{d_i} \le \frac{c_{i+1}}{d_{i+1}}, \qquad i = 1, \ldots, n-1,$$

we have

$$a_{ij} = c_{\min\{i,j\}} d_{\max\{i,j\}} = \min\left\{\frac{c_i}{d_i}, \frac{c_j}{d_j}\right\} d_i d_j.$$

Let

$$b_i = \frac{c_i}{d_i}.$$

Then $0 < b_1 \le \cdots \le b_n$. Setting

$$B = \left(b_{\min\{i,j\}}\right)_{i,j=1}^{n}$$

we have

$$A\begin{pmatrix} i_1, \ldots, i_r \\ j_1, \ldots, j_r \end{pmatrix} = d_{i_1} \cdots d_{i_r} d_{j_1} \cdots d_{j_r} B \begin{pmatrix} i_1, \ldots, i_r \\ j_1, \ldots, j_r \end{pmatrix}.$$

From Proposition 4.1 B is totally positive if $0 < b_1 < \cdots < b_n$. By continuity B is totally positive for $0 \le b_1 \le \cdots \le b_n$. Thus A is totally positive. In fact, from the previous analysis it follows that

$$A\begin{pmatrix} i_1, \ldots, i_r \\ j_1, \ldots, j_r \end{pmatrix} = 0$$

unless

$$i_1, j_1 < i_2, j_2 < \cdots < i_r, j_r$$

in which case we have

$$\begin{aligned} A\begin{pmatrix} i_1, \ldots, i_r \\ j_1, \ldots, j_r \end{pmatrix} &= d_{i_1} \cdots d_{i_r} d_{j_1} \cdots d_{j_r} \frac{c_{\alpha_1}}{d_{\alpha_1}} \prod_{k=1}^{r-1} \left(\frac{c_{\alpha_{k+1}}}{d_{\alpha_{k+1}}} - \frac{c_{\beta_k}}{d_{\beta_k}} \right) \\ &= c_{\alpha_1} \prod_{k=1}^{r-1} \begin{vmatrix} c_{\alpha_{k+1}} & c_{\beta_k} \\ d_{\alpha_{k+1}} & d_{\beta_k} \end{vmatrix} d_{\beta_r}, \end{aligned}$$

where the α_k and β_k are as defined in Proposition 4.1. □

4.5 Jacobi matrices

A *Jacobi matrix* is the term given to a square tridiagonal matrix, i.e., any matrix $A = (a_{ij})_{i,j=1}^{n}$ where $a_{ij} = 0$ if $|i - j| \ge 2$. In other words, it is an

$n \times n$ matrix of the form

$$A = \begin{pmatrix} a_1 & b_1 & 0 & \cdots & 0 & 0 \\ c_1 & a_2 & b_2 & \cdots & 0 & 0 \\ 0 & c_2 & a_3 & \cdots & 0 & 0 \\ \vdots & \vdots & \vdots & \ddots & \vdots & \vdots \\ 0 & 0 & 0 & \cdots & a_{n-1} & b_{n-1} \\ 0 & 0 & 0 & \cdots & c_{n-1} & a_n \end{pmatrix}$$

where $a_{ii} = a_i$, $i = 1, \ldots, n$, and $a_{i,i+1} = b_i$, $a_{i+1,i} = c_i$, $i = 1, \ldots, n-1$. From this form it is readily verified that

$$A\begin{pmatrix} i_1, \ldots, i_r \\ j_1, \ldots, j_r \end{pmatrix} = 0$$

if $|i_k - j_k| \geq 2$ for any $k \in \{1, \ldots, r\}$, and if $|i_\ell - j_\ell| = 1$ then

$$A\begin{pmatrix} i_1, \ldots, i_r \\ j_1, \ldots, j_r \end{pmatrix} = A\begin{pmatrix} i_1, \ldots, i_{\ell-1} \\ j_1, \ldots, j_{\ell-1} \end{pmatrix} A\begin{pmatrix} i_\ell \\ j_\ell \end{pmatrix} A\begin{pmatrix} i_{\ell+1}, \ldots, i_r \\ j_{\ell+1}, \ldots, j_r \end{pmatrix}.$$

From these formulæ we easily see how to calculate every minor of A. Namely, if

$$i_1 = j_1, \ldots, i_{k_1} = j_{k_1}, i_{k_1+1} \neq j_{k_1+1}, \ldots, i_{k_2} \neq j_{k_2},$$
$$i_{k_2+1} = j_{k_2+1}, \ldots, i_{k_3} = j_{k_3}, \ldots$$

then

$$A\begin{pmatrix} i_1, \ldots, i_r \\ j_1, \ldots, j_r \end{pmatrix}$$
$$= A\begin{pmatrix} i_1, \ldots, i_{k_1} \\ j_1, \ldots, j_{k_1} \end{pmatrix} A\begin{pmatrix} i_{k_1+1} \\ j_{k_1+1} \end{pmatrix} \cdots A\begin{pmatrix} i_{k_2} \\ j_{k_2} \end{pmatrix} A\begin{pmatrix} i_{k_2+1}, \ldots, i_{k_3} \\ j_{k_2+1}, \ldots, j_{k_3} \end{pmatrix} \cdots.$$

Furthermore, when calculating a principal minor

$$A\begin{pmatrix} i_1, \ldots, i_r \\ i_1, \ldots, i_r \end{pmatrix}$$

we see that if $i_j + 1 < i_{j+1}$ for some $j \in \{1, \ldots, r-1\}$, then

$$A\begin{pmatrix} i_1, \ldots, i_r \\ i_1, \ldots, i_r \end{pmatrix} = A\begin{pmatrix} i_1, \ldots, i_j \\ i_1, \ldots, i_j \end{pmatrix} A\begin{pmatrix} i_{j+1}, \ldots, i_r \\ i_{j+1}, \ldots, i_r \end{pmatrix}.$$

Thus we have the following result.

Theorem 4.3 *A Jacobi matrix A is totally positive if and only if all its off-diagonal elements $\{b_i\}$, $\{c_i\}$ and all its principal minors containing consecutive rows and columns are nonnegative.*

There are two additional facts worth noting concerning totally positive Jacobi matrices. In our explanation thereof we will use the following formula based on the Laplace expansion by minors. Namely, for a Jacobi matrix A and any $i \in \{1, \ldots, n-1\}$ we have

$$A\begin{pmatrix}1,\ldots,n\\1,\ldots,n\end{pmatrix} = A\begin{pmatrix}1,\ldots,i\\1,\ldots,i\end{pmatrix} A\begin{pmatrix}i+1,\ldots,n\\i+1,\ldots,n\end{pmatrix} - a_{i,i+1}a_{i+1,i}A\begin{pmatrix}1,\ldots,i-1\\1,\ldots,i-1\end{pmatrix} A\begin{pmatrix}i+2,\ldots,n\\i+2,\ldots,n\end{pmatrix}. \quad (4.1)$$

The first fact we will explain is that the inverse of a nonsingular symmetric Jacobi matrix is a Green's matrix. The reason for this is the following. Assume A is an $n \times n$ symmetric Jacobi matrix. Let $\{c_i\}$ denote the off-diagonal entries of A. Assume, for convenience, that each of the c_i is nonzero. Set $B = A^{-1}$. Then

$$b_{ij} = \frac{(-1)^{i+j} A\begin{pmatrix}1,\ldots,\widehat{j},\ldots,n\\1,\ldots,\widehat{i},\ldots,n\end{pmatrix}}{A\begin{pmatrix}1,\ldots,n\\1,\ldots,n\end{pmatrix}}.$$

Assuming $i < j$ this reduces to

$$b_{ij} = \frac{(-1)^{i+j} A\begin{pmatrix}1,\ldots,i-1\\1,\ldots,i-1\end{pmatrix} c_i \cdots c_{j-1} A\begin{pmatrix}j+1,\ldots,n\\j+1,\ldots,n\end{pmatrix}}{A\begin{pmatrix}1,\ldots,n\\1,\ldots,n\end{pmatrix}}.$$

Similar formulæ hold in the case $i = j$ and $i > j$. Set

$$d_i = \frac{(-1)^i A\begin{pmatrix}1,\ldots,i-1\\1,\ldots,i-1\end{pmatrix} c_i \cdots c_{n-1}}{A\begin{pmatrix}1,\ldots,n\\1,\ldots,n\end{pmatrix}}$$

and

$$e_j = (-1)^j \frac{A\begin{pmatrix}j+1,\ldots,n\\j+1,\ldots,n\end{pmatrix}}{c_j \cdots c_{n-1}}.$$

Then it is readily verified that

$$b_{ij} = d_{\min\{i,j\}} e_{\max\{i,j\}}.$$

If A is a nonsingular symmetric totally positive Jacobi matrix and $B = A^{-1}$, then the matrix

$$((-1)^{i+j} b_{ij})$$

is a nonsingular totally positive Green's matrix, i.e., setting $\widetilde{d}_i = (-1)^i d_i$ and $\widetilde{e}_i = (-1)^i e_i$ we have that the $\widetilde{d}_1, \ldots, \widetilde{d}_n$ and $\widetilde{e}_1, \ldots, \widetilde{e}_n$ are all positive,

$$(-1)^{i+j} b_{ij} = \widetilde{d}_{\min\{i,j\}} \widetilde{e}_{\max\{i,j\}},$$

and

$$\frac{\widetilde{d}_i}{\widetilde{e}_i} < \frac{\widetilde{d}_{i+1}}{\widetilde{e}_{i+1}}, \qquad i = 1, \ldots, n-1.$$

From the above definitions of the $\{\widetilde{d}_i\}$ and $\{\widetilde{e}_i\}$ this latter inequality is equivalent to proving

$$\frac{A\begin{pmatrix}1,\ldots,i-1\\1,\ldots,i-1\end{pmatrix}}{A\begin{pmatrix}i+1,\ldots,n\\i+1,\ldots,n\end{pmatrix}} c_i^2 < \frac{A\begin{pmatrix}1,\ldots,i\\1,\ldots,i\end{pmatrix}}{A\begin{pmatrix}i+2,\ldots,n\\i+2,\ldots,n\end{pmatrix}}, \qquad i = 1, \ldots, n-1,$$

i.e.,

$$0 < A\begin{pmatrix}1,\ldots,i\\1,\ldots,i\end{pmatrix} A\begin{pmatrix}i+1,\ldots,n\\i+1,\ldots,n\end{pmatrix} - c_i^2 A\begin{pmatrix}1,\ldots,i-1\\1,\ldots,i-1\end{pmatrix} A\begin{pmatrix}i+2,\ldots,n\\i+2,\ldots,n\end{pmatrix}.$$

As $c_i = a_{i,i+1} = a_{i+1,i}$, from (4.1) we see that the right-hand-side equals

$$A\begin{pmatrix}1,\ldots,n\\1,\ldots,n\end{pmatrix}$$

which is strictly positive.

The other fact we wish to remark upon are necessary and sufficient conditions for when A is a nonsingular totally positive Jacobi matrix. From Theorem 4.3 we have that A is totally positive if and only if all its off-diagonal elements $\{b_i\}$, $\{c_i\}$ and all its principal minors containing consecutive rows and columns are nonnegative. This latter condition can be further simplified. Namely a nonsingular Jacobi matrix A is totally positive if and only if all its off-diagonal elements $\{b_i\}$, $\{c_i\}$ are nonnegative and all its principal minors composed of initial consecutive rows and columns are strictly positive. This fact follows from Theorem 4.3 and from the identity (4.1) with arbitrary $r \in \{2, \ldots, n\}$ and $s \in \{1, \ldots, r-1\}$. Namely

$$\begin{aligned} A\begin{pmatrix}1,\ldots,r\\1,\ldots,r\end{pmatrix} &= A\begin{pmatrix}1,\ldots,s\\1,\ldots,s\end{pmatrix} A\begin{pmatrix}s+1,\ldots,r\\s+1,\ldots,r\end{pmatrix} \\ &\quad - A\begin{pmatrix}1,\ldots,s-1\\1,\ldots,s-1\end{pmatrix} a_{s,s+1}a_{s+1,s} A\begin{pmatrix}s+2,\ldots,r\\s+2,\ldots,r\end{pmatrix}. \end{aligned}$$

The details of the proof based on this formula are left to the reader.

4.6 Hankel matrices

Assume we are given a sequence $b_0, b_1, \ldots, b_{2n}$. The matrix

$$A = (b_{i+j})_{i,j=0}^{n}$$

is called a *Hankel matrix*, i.e.,

$$A = \begin{pmatrix} b_0 & b_1 & \cdots & b_n \\ b_1 & b_2 & \cdots & b_{n+1} \\ \vdots & \vdots & \ddots & \vdots \\ b_n & b_{n+1} & \cdots & b_{2n} \end{pmatrix}.$$

The following characterizes strictly totally positive Hankel matrices.

Theorem 4.4 *The above Hankel matrix A is strictly totally positive if and only if*

$$\sum_{i,j=0}^{n} b_{i+j} x_i x_j, \qquad \sum_{i,j=0}^{n-1} b_{i+j+1} x_i x_j$$

are both strictly positive definite forms, i.e., the symmetric matrices

$$\begin{pmatrix} b_0 & b_1 & \cdots & b_n \\ b_1 & b_2 & \cdots & b_{n+1} \\ \vdots & \vdots & \ddots & \vdots \\ b_n & b_{n+1} & \cdots & b_{2n} \end{pmatrix}$$

and

$$\begin{pmatrix} b_1 & b_2 & \cdots & b_n \\ b_2 & b_3 & \cdots & b_{n+1} \\ \vdots & \vdots & \ddots & \vdots \\ b_n & b_{n+1} & \cdots & b_{2n-1} \end{pmatrix}$$

are strictly positive definite.

Proof Recall that one equivalent definition of strict positive definiteness for symmetric matrices is the strict positivity of the initial principal minors. Thus the necessity of the above conditions is obvious since every symmetric strictly totally positive matrix must be strictly positive definite. It remains to prove the sufficiency.

Let us therefore assume that the two matrices in the statement of the theorem are strictly positive definite. Let $p(t) = \sum_{i=0}^{2n} a_i t^i$ be any (nontrivial) polynomial of degree at most $2n$ that is nonnegative on $[0, \infty)$. We claim that we must have

$$\sum_{i=0}^{2n} a_i b_i > 0.$$

It is known that every polynomial of degree at most $2n$ that is nonnegative on $[0, \infty)$ can be written in the form

$$p(t) = (q(t))^2 + t(r(t))^2$$

where q and r are polynomials of degree at most n and $n - 1$, respectively. Let

$$q(t) = \sum_{j=0}^{n} \mu_j t^j, \qquad r(t) = \sum_{j=0}^{n-1} \sigma_j t^j.$$

Then the condition

$$\sum_{i=0}^{2n} a_i b_i > 0$$

whenever $p(t)$ is a nonnegative (nontrivial) polynomial on $[0, \infty)$ is easily shown to be equivalent to

$$\sum_{j,k=0}^{n} \mu_j \mu_k b_{j+k} > 0$$

and

$$\sum_{j,k=0}^{n-1} \sigma_j \sigma_k b_{j+k+1} > 0$$

for every choice of (nontrivial) $(\mu_0, \ldots, \mu_n)$ and $(\sigma_0, \ldots, \sigma_{n-1})$. These latter conditions are exactly the strict positive definiteness of our two matrices.

A minor digression is now in order. The study of moments was a central motivating theme in the development of functional analysis. An important precursor in this development is Stieltjes [1894–95]. In this paper Stieltjes discussed the problem of necessary and sufficient conditions on a sequence $(b_i)_{i=0}^{\infty}$ so that the $(b_i)_{i=0}^{\infty}$ can be represented in the form

$$b_i = \int_0^{\infty} t^i \, d\alpha(t), \qquad i = 0, 1, 2, \ldots$$

for some nonnegative Borel measure $d\alpha$. (We have rephrased Stieltjes' original problem as there were then no Borel measures. In fact the Stieltjes integral was introduced in that paper explicitly to deal with this problem.) We now know that this is essentially equivalent (via the Hahn–Banach Theorem) to asking for the existence of a continuous nonnegative linear functional L on $C[0,\infty)$ satisfying

$$L(t^i) = b_i, \qquad i = 0, 1, 2, \ldots$$

In the finite case things are somewhat simpler. We have

$$b_i = \int_0^\infty t^i d\alpha(t), \qquad i = 0, 1, \ldots, 2n,$$

(and the $(b_i)_{i=0}^{2n}$ are interior points of the associated convex moment cone) if and only if

$$\sum_{i=0}^{2n} a_i b_i > 0$$

whenever $p(t) = \sum_{i=0}^{2n} a_i t^i$ is a nonnegative (nontrivial) polynomial on $[0, \infty)$.

It also follows from the theory of moments that each such $(b_0, \ldots, b_{2n})$ can be represented in the form

$$b_i = \sum_{j=1}^{n+1} \lambda_j \xi_j^i, \qquad i = 0, 1, \ldots, 2n, \tag{4.2}$$

where $\lambda_j > 0$, $j = 1, \ldots, n+1$, and $0 < \xi_1 < \cdots < \xi_{n+1}$.

Let us therefore assume that the $\{b_i\}$ are as given in (4.2). Then, as is readily verified,

$$\begin{pmatrix} b_0 & \ldots & b_n \\ \vdots & \ddots & \vdots \\ b_n & \ldots & b_{2n} \end{pmatrix} =$$

$$\begin{pmatrix} \xi_1^0 & \ldots & \xi_{n+1}^0 \\ \vdots & \ddots & \vdots \\ \xi_1^n & \ldots & \xi_{n+1}^n \end{pmatrix} \begin{pmatrix} \lambda_1 & \ldots & 0 \\ \vdots & \ddots & \vdots \\ 0 & \ldots & \lambda_{n+1} \end{pmatrix} \begin{pmatrix} \xi_1^0 & \ldots & \xi_1^n \\ \vdots & \ddots & \vdots \\ \xi_{n+1}^0 & \ldots & \xi_{n+1}^n \end{pmatrix}.$$

Set

$$X = \begin{pmatrix} \xi_1^0 & \ldots & \xi_{n+1}^0 \\ \vdots & \ddots & \vdots \\ \xi_1^n & \ldots & \xi_{n+1}^n \end{pmatrix}$$

and

$$\Lambda = \begin{pmatrix} \lambda_1 & \dots & 0 \\ \vdots & \ddots & \vdots \\ 0 & \dots & \lambda_{n+1} \end{pmatrix}.$$

Thus

$$A = X\Lambda X^T.$$

As X and X^T are strictly totally positive matrices (X is a Vandermonde matrix) and Λ is a diagonal nonsingular totally positive matrix, the result follows. That is,

$$A\begin{pmatrix} i_1, \dots, i_r \\ j_1, \dots, j_r \end{pmatrix} =$$

$$\sum_{1 \le k_1 < \cdots < k_r \le n+1} \lambda_{k_1} \cdots \lambda_{k_r} X\begin{pmatrix} i_1, \dots, i_r \\ k_1, \dots, k_r \end{pmatrix} X\begin{pmatrix} j_1, \dots, j_r \\ k_1, \dots, k_r \end{pmatrix} > 0. \quad \square$$

4.7 Toeplitz matrices

Assuming we are given a bi-infinite sequence

$$\dots, a_{-2}, a_{-1}, a_0, a_1, a_2, \dots,$$

the associated *Toeplitz matrix* is defined by

$$A = (a_{j-i})_{i,j=1}^{\infty}.$$

If we are given a one-sided infinite sequence

$$a_0, a_1, a_2, \dots,$$

then we understand this to mean that $a_{-n} = 0$, $n = 1, 2, \dots$, in the above definition. Sequences that give rise to totally positive Toeplitz matrices have been totally characterized in terms of their generating functions, i.e., representations of $\sum_{k=-\infty}^{\infty} a_k z^k$ or $\sum_{k=0}^{\infty} a_k z^k$.

In the latter case, with the normalization $a_0 = 1$, the sequence

$$a_0, a_1, a_2, \dots,$$

gives rise to a totally positive Toeplitz matrix

$$A = (a_{j-i})_{i,j=1}^{\infty}$$

if and only if $\sum_{k=0}^{\infty} a_k z^k$ has the form

$$e^{\gamma z}\frac{\prod_{i=1}^{\infty}(1+\alpha_i z)}{\prod_{i=1}^{\infty}(1-\beta_i z)}$$

where $\gamma \geq 0$, $\alpha_i \geq 0$, $\beta_i \geq 0$, and $\sum_{i=1}^{\infty}(\alpha_i + \beta_i) < \infty$. The proof of this and the corresponding representation for the bi-infinite sequence will not be presented here. However, we will prove the characterization in the simpler case where the sequence has only a finite number of nonzero terms.

Theorem 4.5 *Let $a_0 > 0$, $a_n \neq 0$ and $a_k = 0$ for $k < 0$ and $k > n$. Then*

$$A = (a_{j-i})_{i,j=1}^{\infty}$$

is a totally positive matrix if and only if

$$p(x) = \sum_{k=0}^{n} a_k x^k$$

has n negative zeros. Furthermore, if A is totally positive then

$$A\begin{pmatrix} i_1, \ldots, i_r \\ j_1, \ldots, j_r \end{pmatrix} > 0$$

if and only if

$$a_{j_k - i_k} > 0, \qquad k = 1, \ldots, r,$$

i.e., $0 \leq j_k - i_k \leq n$, $k = 1, \ldots, r$.

Proof Let us first assume that p has n negative zeros. Then, assuming $a_0 = 1$, we have

$$p(x) = \prod_{r=1}^{n}(1 + \alpha_r x)$$

where $\alpha_r > 0$ for $r = 1, \ldots, n$. Let

$$E_r = (b_{j-i}^{(r)})_{i,j=1}^{\infty}$$

where $b_0^{(r)} = 1$, $b_1^{(r)} = \alpha_r$ and $b_k^{(r)} = 0$ for $k \neq 0, 1$. It follows by inspection, since $\alpha_r > 0$, that each E_r is totally positive. Furthermore

$$A = \prod_{r=1}^{n} E_r.$$

This latter result follows from the fact that the generating function of a product of Toeplitz matrices is the product of their generating functions. As A is a product of totally positive matrices, then A is totally positive.

Let us now assume that A is totally positive. We will prove that

$$p(x) = \sum_{k=0}^{n} a_k x^k$$

has n negative zeros. We prove this by contradiction using the variation diminishing property of totally positive matrices. From the fact that A is totally positive (thus $a_k \geq 0$) and $a_0 > 0$, it follows that p has no nonnegative real zeros. Let us assume that p has nonreal zeros. Since the coefficients of p are real, if $p(w) = 0$ then $p(\overline{w}) = 0$. Thus we may assume that w is a zero of p satisfying

$$w = re^{i\theta}$$

where $0 < \theta < \pi$.

Now, as

$$\sum_{k=0}^{n} a_k w^{k+\ell} = 0, \qquad \ell \in \mathbb{Z}$$

we have

$$\sum_{k=0}^{n} a_k r^{k+\ell} \sin(k+\ell)\theta = 0, \qquad \ell \in \mathbb{Z}.$$

Let $\mathbf{x} = (x_0, x_1, x_2, \ldots)$ where

$$x_k = r^k \sin k\theta.$$

Thus

$$A\mathbf{x} = \mathbf{0}.$$

Let $A_{[m]}$ denote the $m \times (n+m)$ submatrix of A obtained by choosing the first m rows and $(n+m)$ columns of A, i.e.,

$$A_{[m]} = \begin{pmatrix} a_0 & \cdots & a_n & & 0 \\ & \ddots & & \ddots & \\ 0 & & a_0 & \cdots & a_n \end{pmatrix}.$$

Set

$$\mathbf{x}_{[m]} = (x_0, \ldots, x_{n+m-1}).$$

Thus

$$A_{[m]}\mathbf{x}_{[m]} = \mathbf{0} \in \mathbb{R}^m.$$

As $n \geq 1$ and because

$$\operatorname{rank} A_{[m]} = m$$

there exists an $\mathbf{x}^* \in \mathbb{R}^{n+m}$ satisfying

$$A_{[m]}\mathbf{x}^* = \mathbf{d}$$

where

$$\mathbf{d} = (1, -1, , \ldots, (-1)^{m-1}) \in \mathbb{R}^m.$$

Thus for any ε

$$A_{[m]}(\mathbf{x}_{[m]} + \varepsilon\mathbf{x}^*) = \varepsilon\mathbf{d}.$$

From the variation diminishing properties of the totally positive matrix $A_{[m]}$ (Theorem 3.4) we have

$$m - 1 = S^-(\varepsilon\mathbf{d}) \leq S^-(\mathbf{x}_{[m]} + \varepsilon\mathbf{x}^*)$$

for any $\varepsilon \neq 0$.

Take $\varepsilon^* \neq 0$ and sufficiently small such that

$$(\mathbf{x}_{[m]} + \varepsilon\mathbf{x}^*)_k(\mathbf{x}_{[m]})_k > 0$$

if $(\mathbf{x}_{[m]})_k = x_k \neq 0$. Note that if $x_k = 0$, $0 < k < n + m - 1$, then since $0 < \theta < \pi$ we have

$$x_{k-1}x_{k+1} < 0.$$

Thus

$$S^-(\mathbf{x}_{[m]} + \varepsilon\mathbf{x}^*) \leq S^-(\mathbf{x}_{[m]}) + 2$$

(because we do have $x_0 = 0$ and can have $x_{n+m-1} = 0$). Therefore

$$m - 1 \leq S^-(\mathbf{x}_{[m]}) + 2,$$

i.e.,

$$m - 3 \leq S^-(\mathbf{x}_{[m]})$$

for all m.

Now, as is easily seen,

$$S^-(\mathbf{x}_{[m]}) \leq \frac{(n+m)\theta}{\pi}$$

simply because $\sin\alpha > 0$ for $0 < \alpha < \pi$ and $\sin\alpha < 0$ for $\pi < \alpha < 2\pi$, and the fact that when going from x_k to x_{k+1} we alter the "argument" of x_k by θ. Thus

$$m - 3 \leq \frac{(n+m)\theta}{\pi}$$

for every m. But

$$\frac{\theta}{\pi} < 1,$$

which implies that the above cannot hold for m sufficiently large. We have arrived at a contradiction.

The last claim of the theorem (namely that A is necessarily an almost strictly totally positive matrix) is fairly elementary. One direction is a consequence of Theorem 1.13, i.e.,

$$A\begin{pmatrix} i_1,\ldots,i_r \\ j_1,\ldots,j_r \end{pmatrix} > 0$$

implies that the diagonal elements of this minor are strictly positive. The converse direction is a simple consequence of Theorem 1.19 and the fact that

$$A\begin{pmatrix} i+1,\ldots,i+r \\ i+1,\ldots,i+r \end{pmatrix} = a_0^r > 0$$

and

$$A\begin{pmatrix} i+1,\ldots,i+r \\ i+1+n,\ldots,i+r+n \end{pmatrix} = a_n^r > 0. \qquad \square$$

The above characterization does not necessarily mean that it is a simple matter to determine whether a given Toeplitz matrix is totally positive or not. It is simply equivalent to the fact that the associated generating polynomial has all negative zeros. We return to this problem in Section 4.9.

Let us consider a few specific examples of totally positive Toeplitz matrices and totally positive matrices derived from Toeplitz matrices.

1. Consider the polynomial

$$p(x) = (1+x)^n.$$

As its zeros are all negative the Toeplitz matrix with

$$a_k = \binom{n}{k}, \qquad k = 0,1,\ldots,n$$

is totally positive.

2. In Theorem 1.11 we proved that if the square matrix A is (strictly) totally positive, then the matrix B defined by

$$b_{1j} = a_{1j}, \qquad \text{for all } j,$$

and for $i \geq 2$

$$b_{ij} = \sum_k b_{i-1,k} a_{kj}, \qquad \text{for all } j,$$

is also (strictly) totally positive. Applying this result to the above Toeplitz matrix where $n = 1$, i.e., $a_0 = a_1 = 1$, $a_m = 0$ if $m \neq 0, 1$, we get

$$b_{11} = b_{12} = 1$$

and for $i \geq 2$

$$b_{i1} = 1, \qquad b_{ij} = b_{i-1,j-1} + b_{i-1,j}, \qquad j \geq 2,$$

which immediately implies that

$$b_{ij} = \binom{i}{j-1}, \qquad i, j = 1, 2, \ldots .$$

Letting $\binom{n}{m} = 0$ for $n < m$ we have that the matrix of binomial coefficients

$$C = \left(\binom{i}{j}\right)_{i,j=0}^{\infty}$$

is totally positive. This is equivalent to the fact that the infinite Toeplitz matrix

$$D = \left(\frac{1}{(i-j)!}\right)_{i,j=0}^{\infty}$$

is totally positive.

3. Applying Theorem 1.11 to the infinite totally positive Toeplitz matrix $A = (a_{ij})_{i,j=0}^{\infty}$, where

$$a_{ij} = \begin{cases} 1, & i \leq j \\ 0, & i > j \end{cases}$$

it easily follows that the matrix

$$B = \left(\binom{i+j}{j}\right)_{i,j=0}^{\infty}$$

is totally positive, which is equivalent to the fact that the Hankel matrix

$$D = ((i+j)!)_{i,j=0}^{\infty}$$

is totally positive. The total positivity of D can also be proven by other methods.

4. Consider the polynomial

$$p(x) = \prod_{i=1}^{n}(1 + q^{i-1}x)$$

for some $q > 0$, $q \neq 1$. These polynomials are called *Gaussian polynomials*. Set

$$p(x) = \sum_{k=0}^{n} a_k x^k$$

and let us calculate the coefficients a_k.

From the fact that

$$(1 + q^n x)p(x) = (1 + x)p(qx)$$

we have

$$(1 + q^n x)\sum_{k=0}^{n} a_k x^k = (1 + x)\sum_{k=0}^{n} a_k q^k x^k.$$

Equating the coefficients of x^k we obtain for $k = 1, \ldots, n$

$$a_k + q^n a_{k-1} = a_k q^k + a_{k-1} q^{k-1}$$

implying

$$a_k = a_{k-1} q^{k-1} \frac{(1 - q^{n-k+1})}{(1 - q^k)}.$$

Now $a_0 = 1$ so that

$$a_k = q^{(k-1)+(k-2)+\cdots+1} \frac{(1 - q^{n-k+1}) \cdots (1 - q^n)}{(1 - q^k) \cdots (1 - q)},$$

which we write as

$$a_k = \begin{bmatrix} n \\ k \end{bmatrix}_q q^{k(k-1)/2}.$$

The

$$\begin{bmatrix} n \\ k \end{bmatrix}_q = \frac{(1 - q^{n-k+1}) \cdots (1 - q^n)}{(1 - q^k) \cdots (1 - q)}$$

are termed q-binomial coefficients.

Thus the Toeplitz matrix with

$$a_k = \begin{bmatrix} n \\ k \end{bmatrix}_q q^{k(k-1)/2}, \qquad k = 0, 1, \ldots, n,$$

is totally positive.

4.8 Generalized Hurwitz matrices

Assume we are given real numbers $a_0, a_1, \ldots, a_n$ where $n \geq M \geq 2$. For ease of exposition, set $a_j = 0$ for $j < 0$ and $j > n$. Let us define

$$P = (p_{ij})_{i,j=1}^{\infty}$$

where

$$p_{ij} = a_{Mj-i}, \qquad i, j = 1, 2, \ldots.$$

Thus

$$P = \begin{pmatrix} a_{M-1} & a_{2M-1} & a_{3M-1} & \cdots \\ a_{M-2} & a_{2M-2} & a_{3M-2} & \cdots \\ \vdots & \vdots & \vdots & \\ a_0 & a_M & a_{2M} & \cdots \\ 0 & a_{M-1} & a_{2M-1} & \cdots \\ 0 & a_{M-2} & a_{2M-2} & \cdots \\ \vdots & \vdots & \vdots & \\ 0 & a_0 & a_M & \cdots \\ 0 & 0 & a_{M-1} & \cdots \\ \vdots & \vdots & \vdots & \end{pmatrix}.$$

The matrix P is not a Toeplitz matrix, but it resembles a Toeplitz matrix. It also contains M Toeplitz submatrices. Simply take all columns and every Mth row.

Somewhat surprisingly we prove that there are a finite number of determinantal conditions that imply that this matrix is totally positive.

Theorem 4.6 *Assume* $a_0 > 0$ *and*

$$P\begin{pmatrix} k, k+1, \ldots, k+r-1 \\ 1, 2, \ldots, r \end{pmatrix} > 0 \tag{4.3}$$

for $k = 1, \ldots, M-1$, *and* $r = 1, \ldots, \lfloor \frac{n+k-1}{M-1} \rfloor$. *Then* P *is totally positive,* $a_k > 0$, $k = 0, 1, \ldots, n$, *and*

$$P\begin{pmatrix} i_1, \ldots, i_r \\ j_1, \ldots, j_r \end{pmatrix} > 0$$

if and only if

$$p_{i_k j_k} = a_{Mj_k - i_k} > 0, \qquad k = 1, \ldots, r,$$

i.e., $0 \leq Mj_k - i_k \leq n$.

Before we prove this theorem note that for $r > \lfloor \frac{n+k-1}{M-1} \rfloor$ the last column of

$$P\begin{bmatrix} k, k+1, \ldots, k+r-1 \\ 1, 2, \ldots, r \end{bmatrix}$$

vanishes identically. That is, the conditions of (4.3), together with $a_0 > 0$, are exactly the demand that all possible nonvanishing minors composed from consecutive rows and initial consecutive columns be strictly positive.

In addition, let $k \in \{1, \ldots, M-1\}$ be such that

$$r = \frac{n+k-1}{M-1}$$

is an integer. Such a k exists and $r \geq 2$. For this k and r

$$P\begin{pmatrix} k, \ldots, k+r-1 \\ 1, \ldots, r \end{pmatrix} = a_n P\begin{pmatrix} k, \ldots, k+r-2 \\ 1, \ldots, r-1 \end{pmatrix}.$$

Thus, from (4.3), it immediately follows that $a_n > 0$.

Our proof is via induction on n. In this proof we use the following result.

Lemma 4.7 *Let*

$$b_{Mj+k-1} = a_{Mj+k}, \qquad k = 1, \ldots, M-1,\ j = 0, 1, \ldots,$$

and

$$b_{Mj-1} = a_{Mj} - \frac{a_0}{a_1} a_{Mj+1}, \qquad j = 1, 2, \ldots,$$

where applicable, to define $b_0, \ldots, b_{n-1}$. Set $b_k = 0$ for $k < 0$ and $k > n-1$. Define

$$Q = (q_{ij})_{i,j=1}^{\infty}$$

where

$$q_{ij} = b_{Mj-i}, \qquad i, j = 1, 2, \ldots.$$

Then Q satisfies the conditions of Theorem 4.6 (with $n-1$ replacing n).

Proof Note that $b_0 = a_1 = p_{M-1,1} > 0$ and by the above

$$b_{-1} = a_0 - \frac{a_0}{a_1} a_1 = 0.$$

Thus $b_j = 0$ for $j < 0$ and, by definition, $b_j = 0$ for $j > n-1$.

Recall that

$$P = \begin{pmatrix} a_{M-1} & a_{2M-1} & a_{3M-1} & \cdots \\ a_{M-2} & a_{2M-2} & a_{3M-2} & \cdots \\ \vdots & \vdots & \vdots & \\ a_0 & a_M & a_{2M} & \cdots \\ 0 & a_{M-1} & a_{2M-1} & \cdots \\ 0 & a_{M-2} & a_{2M-2} & \cdots \\ \vdots & \vdots & \vdots & \\ 0 & a_0 & a_M & \cdots \\ 0 & 0 & a_{M-1} & \cdots \\ \vdots & \vdots & \vdots & \end{pmatrix}$$

and Q has exactly this same form, i.e.,

$$Q = \begin{pmatrix} b_{M-1} & b_{2M-1} & b_{3M-1} & \cdots \\ b_{M-2} & b_{2M-2} & b_{3M-2} & \cdots \\ \vdots & \vdots & \vdots & \\ b_0 & b_M & b_{2M} & \cdots \\ 0 & b_{M-1} & b_{2M-1} & \cdots \\ 0 & b_{M-2} & b_{2M-2} & \cdots \\ \vdots & \vdots & \vdots & \\ 0 & b_0 & b_M & \cdots \\ 0 & 0 & b_{M-1} & \cdots \\ \vdots & \vdots & \vdots & \end{pmatrix}.$$

For $k = 2, \ldots, M-1$, it is easily verified that

$$Q\begin{pmatrix} k, k+1, \ldots, k+r-1 \\ 1, 2, \ldots, r \end{pmatrix} = P\begin{pmatrix} k-1, k, \ldots, k+r-2 \\ 1, 2, \ldots, r \end{pmatrix} > 0$$

for $r = 1, \ldots, \lfloor \frac{n+(k-1)-1}{M-1} \rfloor = \lfloor \frac{(n-1)+k-1}{M-1} \rfloor$. We must also consider

$$Q\begin{pmatrix} 1, 2, \ldots, r \\ 1, 2, \ldots, r \end{pmatrix}.$$

Now, for $2 \le r \le \lfloor \frac{n+M-2}{M-1} \rfloor$,

$$\begin{aligned}
0 &< P\begin{pmatrix} M-1, M, \ldots, M+r-2 \\ 1, 2, \ldots, r \end{pmatrix} = \begin{vmatrix} a_1 & a_{M+1} & a_{2m+1} & \cdots \\ a_0 & a_M & a_{2M} & \cdots \\ 0 & a_{M-1} & a_{2M-1} & \cdots \\ \vdots & \vdots & \vdots & \ddots \end{vmatrix} \\
&= a_1 \begin{vmatrix} a_M & a_{2M} & \cdots \\ a_{M-1} & a_{2M-1} & \cdots \\ \vdots & \vdots & \ddots \end{vmatrix} - a_0 \begin{vmatrix} a_{M+1} & a_{2M+1} & \cdots \\ a_{M-1} & a_{2M-1} & \cdots \\ \vdots & \vdots & \ddots \end{vmatrix} \\
&= a_1 \begin{vmatrix} a_M - \frac{a_0}{a_1} a_{M+1} & a_{2M} - \frac{a_0}{a_1} a_{2M+1} & \cdots \\ a_{M-1} & a_{2M-1} & \cdots \\ \vdots & \vdots & \ddots \end{vmatrix} \\
&= a_1 Q\begin{pmatrix} 1, 2, \ldots, r-1 \\ 1, 2, \ldots, r-1 \end{pmatrix}.
\end{aligned}$$

As $a_1 > 0$ we have

$$Q\begin{pmatrix} 1, 2, \ldots, r-1 \\ 1, 2, \ldots, r-1 \end{pmatrix} > 0$$

for $1 \le r-1 \le \lfloor \frac{n-1}{M-1} \rfloor$, which is exactly what we wished to prove. □

We now proceed to the proof of our theorem.

Proof of Theorem 4.6 We start with the case $n = M$. Here

$$P = \begin{pmatrix} a_{M-1} & 0 & \cdots \\ a_{M-2} & 0 & \cdots \\ \vdots & \vdots & \\ a_1 & 0 & \cdots \\ a_0 & a_M & \cdots \\ 0 & a_{M-1} & \cdots \\ \vdots & \vdots & \vdots \\ 0 & a_0 & \cdots \\ 0 & 0 & \cdots \\ \vdots & \vdots & \end{pmatrix}.$$

From (4.3) we have $p_{k1} = a_{M-k} > 0$, $k = 1, \ldots, M-1$, and

$$0 < P\begin{pmatrix} M-1, M \\ 1, 2 \end{pmatrix} = \begin{vmatrix} a_1 & 0 \\ a_0 & a_M \end{vmatrix} = a_1 a_M.$$

Thus $a_1, a_1, \ldots, a_M > 0$. The other claims follow easily from this fact and the geometric form of P.

We now apply induction assuming the result is valid for $n-1$, where $n > M$. We start with the $a_0, \ldots, a_n$ and P as in the statement of Theorem 4.6. From Lemma 4.7, the $b_0, \ldots, b_{n-1}$ and the Q therein satisfy (4.3) for $n-1$, and therefore Q is totally positive, $b_k > 0$, $k = 0, \ldots, n-1$, and

$$Q\begin{pmatrix} i_1, \ldots, i_r \\ j_1, \ldots, j_r \end{pmatrix} > 0$$

if and only if

$$q_{i_k j_k} = b_{Mj_k - i_k} > 0, \qquad k = 1, \ldots, r,$$

i.e., $0 \le Mj_k - i_k \le n-1$.

From Lemma 4.7 we have

$$\begin{aligned} a_{Mj+k} &= b_{Mj+k-1}, \qquad k = 1, \ldots, M-1 \\ a_{Mj} &= b_{Mj-1} + cb_{Mj}, \qquad k = 0, \end{aligned} \tag{4.4}$$

where $c > 0$. (Specifically $c = a_0/a_1$ and we are given $a_0 > 0$ and $a_1 = p_{M-1,1} > 0$.) Thus P is obtained from Q by an addition of a positive multiple of one row to a succeeding row and shifting (renumbering) of the rows. Both these operations (see Proposition 1.5) preserve totally positive. Thus P is totally positive. The fact that $a_0, a_1, \ldots, a_n > 0$ also follows easily. We are given $a_0 > 0$, and the positivity of the other a_k is a consequence of the positivity of $b_1, b_1, \ldots, b_{n-1}$ and (4.4).

For any nonsingular totally positive matrix the diagonal elements are positive (see Theorem 1.13). That is, if

$$P\begin{pmatrix} i_1, \ldots, i_r \\ j_1, \ldots, j_r \end{pmatrix} > 0$$

then

$$p_{i_k j_k} = a_{Mj_k - i_k} > 0, \qquad k = 1, \ldots, r,$$

i.e., $0 \le Mj_k - i_k \le n$. It remains to prove the converse. As such let us assume that

$$0 \le Mj_k - i_k \le n, \qquad k = 1, \ldots, r.$$

We wish to prove that

$$P\begin{pmatrix} i_1, \ldots, i_r \\ j_1, \ldots, j_r \end{pmatrix} > 0.$$

Based on (4.4) we can write

$$p_{i_s j_t} = a_{Mj_t - i_s} = b_{Mj_t - i_s - 1} + c(i_s) b_{Mj_t - i_s}$$

where

$$c(i_s) = \begin{cases} 0, & i_s \text{ is not a multiple of } M \\ c, & i_s \text{ is a multiple of } M. \end{cases}$$

Now

$$P\begin{pmatrix} i_1, \ldots, i_r \\ j_1, \ldots, j_r \end{pmatrix} = \det (p_{i_s j_t}) = \det (b_{Mj_t - i_s - 1} + c(i_s) b_{Mj_t - i_s}).$$

Thus the desired minor of P is the nonnegative sum of various minors of Q. All the minors of Q are nonnegative. Our task is to show that at least one of the minors in this sum is positive, and has positive coefficient. This we do by defining

$$i'_s = i_s + 1 \quad \text{if } 0 < Mj_s - i_s \le n$$

and

$$i'_s = i_s \quad \text{if } Mj_s - i_s = 0.$$

(Note that in the latter case $c(i'_s) = c(i_s) = c > 0$.)

As

$$0 \le Mj_k - i_k \le n, \qquad k = 1, \ldots, r,$$

then from the above construction

$$0 \le Mj_k - i'_k \le n - 1, \qquad k = 1, \ldots, r.$$

Furthermore

$$i'_1 < \cdots < i'_r.$$

To see this note that if $i'_s = i_s + 1$ then obviously $i'_{s-1} < i'_s$, while if $i'_s = i_s$ then $Mj_s = i_s$ and

$$i'_{s-1} \le Mj_{s-1} < Mj_s = i'_s.$$

Therefore

$$Q\begin{pmatrix} i'_1, \ldots, i'_r \\ j_1, \ldots, j_r \end{pmatrix} > 0.$$

This minor, with positive coefficient, is one of the minors involved in calculating the desired minor of P. Thus

$$P\begin{pmatrix} i_1, \ldots, i_r \\ j_1, \ldots, j_r \end{pmatrix} > 0,$$

which proves the theorem. □

Remark The matrix

$$P = \begin{pmatrix} a_1 & a_3 & a_5 & \cdots \\ a_0 & a_2 & a_4 & \cdots \\ 0 & a_1 & a_3 & \cdots \\ 0 & a_0 & a_2 & \cdots \\ 0 & 0 & a_1 & \cdots \\ \vdots & \vdots & \vdots & \ddots \end{pmatrix}$$

is called a *Hurwitz matrix*. This is the matrix of Theorem 4.6 with $M = 2$. The conditions (4.3) of Theorem 4.6 in this case reduce to

$$P\begin{pmatrix} 1,\ldots,r \\ 1,\ldots,r \end{pmatrix} > 0, \qquad r = 1,\ldots,n.$$

These inequalities, together with $a_0 > 0$, are equivalent to the fact that the polynomial

$$p(x) = a_0x^n + a_1x^{n-1} + \cdots + a_n$$

is a *Hurwitz polynomial*, i.e., has all its zeros in the open left-hand plane $\mathrm{Re}\,(z) < 0$. If p has all its zeros in the closed left-hand plane $\mathrm{Re}\,(z) \le 0$, then P is totally positive. Unfortunately the converse is not true.

4.9 More on Toeplitz matrices

In Section 4.7 we considered infinite Toeplitz matrices with a finite number of nonzero terms and proved that the matrix is totally positive if and only if its associated generating polynomial has only negative zeros. In other words, we replaced one problem by another one which, while providing an insight into the theory, does not always constitute a useful criterion for determining whether a given matrix is in fact totally positive. In this short section we apply the theory of Hurwitz matrices in order to generate verifiable conditions for when such a matrix is totally positive, i.e., when a polynomial has only negative zeros.

We present most of the results here without proof. Their proofs would take us too far afield. References can be found in the last section of this chapter.

Let

$$p(x) = \sum_{k=0}^{n} a_k x^{n-k}$$

with $a_0 > 0$, and set

$$q(x) = p(x^2) + xp'(x^2).$$

Then we have

Theorem 4.8 *The polynomial q is a Hurwitz polynomial if and only if p has simple negative zeros.*

Let

$$q(x) = \sum_{k=0}^{2n} b_k x^{2n-k}.$$

Then

$$b_{2k} = a_k, \qquad k = 0, 1, \ldots, n,$$

and

$$b_{2k+1} = (n - k)a_k, \qquad k = 1, \ldots, n - 1.$$

Let A be the infinite Toeplitz matrix based on p, i.e.,

$$A = \begin{pmatrix} a_0 & a_1 & a_2 & \cdots & a_n & 0 & 0 & \cdots \\ 0 & a_0 & a_1 & \cdots & a_{n-1} & a_n & 0 & \cdots \\ 0 & 0 & a_0 & \cdots & a_{n-2} & a_{n-1} & a_n & \cdots \\ \vdots & \vdots & \vdots & \ddots & \vdots & \vdots & \vdots & \ddots \end{pmatrix}$$

and B be the infinite Hurwitz matrix based on q, i.e.,

$$B = \begin{pmatrix} b_1 & b_3 & b_5 & \cdots \\ b_0 & b_2 & b_4 & \cdots \\ 0 & b_1 & b_3 & \cdots \\ 0 & b_0 & b_2 & \cdots \\ 0 & 0 & b_1 & \cdots \\ \vdots & \vdots & \vdots & \ddots \end{pmatrix} = \begin{pmatrix} na_0 & (n-1)a_1 & (n-2)a_2 & \cdots \\ a_0 & a_1 & a_2 & \cdots \\ 0 & na_0 & (n-1)a_1 & \cdots \\ 0 & a_0 & a_1 & \cdots \\ 0 & 0 & na_0 & \cdots \\ \vdots & \vdots & \vdots & \ddots \end{pmatrix}.$$

Theorem 4.9 *The following are equivalent:*

(i) *A is totally positive.*
(ii) *B is totally positive.*
(iii) *p has n negative zeros.*

Proof From Theorem 4.5, conditions (i) and (iii) are equivalent. Furthermore, if B is totally positive, then since A is a submatrix of B it follows that A is totally positive.

Assume p has n negative zeros. Perturb these zeros so that they are simple and negative. Then from Theorem 4.8 q is a Hurwitz polynomial and thus the associated Hurwitz matrix B is totally positive. Perturb back.

□

From the results of Section 4.8, $b_0 = a_0 > 0$ and

$$B\begin{pmatrix}1,\ldots,k\\1,\ldots,k\end{pmatrix} > 0, \qquad k = 1,\ldots,2n,$$

is equivalent to the fact that q is a Hurwitz polynomial. Thus we have the following.

Theorem 4.10 *The polynomial*

$$p(x) = \sum_{k=0}^{n} a_k x^{n-k}$$

with $a_0 > 0$ has n simple negative zeros if and only if

$$B\begin{pmatrix}1,\ldots,k\\1,\ldots,k\end{pmatrix} > 0, \qquad k = 1,\ldots,2n.$$

When considering the total positivity of the associated Toeplitz matrix, the necessity of

$$a_k > 0, \qquad k = 0, 1, \ldots, n,$$

is obvious. With this a priori assumption it suffices to verify the positivity of half of the above determinants.

Proposition 4.11 *Let p be as above and assume that $a_k > 0$, $k = 0, 1, \ldots, n$. Then p has n simple negative zeros if and only if*

$$B\begin{pmatrix}1,\ldots,2k+1\\1,\ldots,2k+1\end{pmatrix} > 0, \qquad k = 1,\ldots,n-1.$$

When p has multiple zeros then q has zeros on the imaginary axis and the situation is more complicated.

4.10 Hadamard products of totally positive matrices

Given $n \times m$ matrices $A = (a_{ij})$ and $B = (b_{ij})$ we define their *Hadamard product* as the $n \times m$ matrix formed by the entrywise product of the elements of A and B, i.e., it is the $n \times m$ matrix $C = (c_{ij})$ where

$$c_{ij} = a_{ij} b_{ij}.$$

We use the notation

$$C = A \circ B.$$

While certain classes of matrices are closed under the operation of Hadamard products, the class of totally positive matrices is not one of them. For example, the matrix

$$A = \begin{pmatrix} 1 & 1 & 1 \\ 1 & 1 & 1 \\ 0 & 1 & 1 \end{pmatrix}$$

is totally positive, while

$$A \circ A^T = \begin{pmatrix} 1 & 1 & 0 \\ 1 & 1 & 1 \\ 0 & 1 & 1 \end{pmatrix}$$

has determinant equal to -1. (As strictly totally positive matrices are dense in the class of totally positive matrices, it follows that there also exist strictly totally positive matrices whose Hadamard product is not totally positive.)

However, there are numerous subclasses of totally positive matrices that are closed under the operation of taking Hadamard products. In this section we detail some of them.

1. Consider a strictly totally positive matrix satisfying the criteria in Section 2.6. That is, let $A = (a_{ij})$ be an $n \times n$ matrix, all of whose entries are strictly positive and where

$$a_{ij}a_{i+1,j+1} > \left(4\cos^2 \frac{\pi}{n+1}\right) a_{i,j+1}a_{i+1,j}$$

for all $i, j = 1, \ldots, n-1$. Now let $B = (b_{ij})$ be any $n \times n$ matrix, all of whose entries are strictly positive, and such that

$$B\begin{pmatrix} i, i+1 \\ j, j+1 \end{pmatrix} \geq 0$$

for all $i, j = 1, \ldots, n-1$. Then from Theorem 2.16 the Hadamard product

$$A \circ B = (a_{ij}b_{ij})$$

is strictly totally positive.

2. Green's matrices. If A and B are totally positive Green's matrices, then $A \circ B$ is also a totally positive Green's matrix. This immediately follows from the definition of a Green's matrix and the characterization of Green matrices as given in Theorem 4.2.

3. Jacobi matrices. We recall that a Jacobi matrix is a square tridiagonal matrix, i.e., a matrix $A = (a_{ij})_{i,j=1}^n$ where $a_{ij} = 0$ if $|i-j| \geq 2$. Furthermore, from Theorem 4.3, a Jacobi matrix is totally positive if and only if all its off-diagonal elements and all its principal minors containing consecutive rows and columns are nonnegative. We prove the following.

Proposition 4.12 *The Hadamard product of two totally positive Jacobi matrices is a totally positive Jacobi matrix.*

Proof Assume we are given the totally positive Jacobi matrix

$$A = \begin{pmatrix} a_{11} & a_{12} & 0 & \cdots & 0 \\ a_{21} & a_{22} & a_{23} & \cdots & 0 \\ 0 & a_{32} & a_{33} & \cdots & 0 \\ \vdots & \vdots & \vdots & \ddots & \vdots \\ 0 & 0 & 0 & \cdots & a_{nn} \end{pmatrix}.$$

Let $c \geq d \geq 0$, $c > 0$. Then, assuming $a_{11} > 0$, the matrix

$$A_{c,d} = \begin{pmatrix} ca_{11} & a_{12} & 0 & \cdots & 0 \\ 0 & a_{22} - \frac{da_{12}a_{21}}{ca_{11}} & a_{23} & \cdots & 0 \\ 0 & a_{32} & a_{33} & \cdots & 0 \\ \vdots & \vdots & \vdots & \ddots & \vdots \\ 0 & 0 & 0 & \cdots & a_{nn} \end{pmatrix}$$

is also a totally positive Jacobi matrix. To see this let

$$\widetilde{A}_{c,d} = \begin{pmatrix} ca_{11} & a_{12} & 0 & \cdots & 0 \\ da_{21} & a_{22} & a_{23} & \cdots & 0 \\ 0 & a_{32} & a_{33} & \cdots & 0 \\ \vdots & \vdots & \vdots & \ddots & \vdots \\ 0 & 0 & 0 & \cdots & a_{nn} \end{pmatrix}.$$

The matrix $\widetilde{A}_{c,d}$ is totally positive since it is obtained from A by multiplying the first column of A by d and adding $(c-d)a_{11}$ to the $(1,1)$ entry. Both operations preserve total positivity (see Proposition 1.5). Now consider $A_{c,d}$. Its off-diagonal elements are nonnegative and

$$A_{c,d}\begin{pmatrix} i,\ldots,j \\ i,\ldots,j \end{pmatrix} = \widetilde{A}_{c,d}\begin{pmatrix} i,\ldots,j \\ i,\ldots,j \end{pmatrix}$$

except when $i=2$, in which case

$$A_{c,d}\begin{pmatrix} 2,\ldots,j \\ 2,\ldots,j \end{pmatrix} = \frac{1}{ca_{11}}\widetilde{A}_{c,d}\begin{pmatrix} 1,\ldots,j \\ 1,\ldots,j \end{pmatrix} \geq 0.$$

Thus $A_{c,d}$ is a totally positive Jacobi matrix.

Let us now assume that A and B are $n \times n$ totally positive Jacobi matrices. Our proof is by induction on n. The result is easily verified for $n=1$ (and $n=2$). Assume the result holds for $r \times r$ matrices, $r \leq n-1$.

Now

$$A \circ B = \begin{pmatrix} a_{11}b_{11} & a_{12}b_{12} & 0 & \cdots & 0 \\ a_{21}b_{21} & a_{22}b_{22} & a_{23}b_{23} & \cdots & 0 \\ 0 & a_{32}b_{32} & a_{33}b_{33} & \cdots & 0 \\ \vdots & \vdots & \vdots & \ddots & \vdots \\ 0 & 0 & 0 & \cdots & a_{nn}b_{nn} \end{pmatrix}.$$

We wish to prove that

$$(A \circ B)\begin{pmatrix} 1,\ldots,n \\ 1,\ldots,n \end{pmatrix} \geq 0$$

to advance the induction. If $a_{11}=0$ then $a_{12}=0$ or $a_{21}=0$, and the above determinant is therefore zero. The similar result holds if $b_{11}=0$. We may therefore assume that $a_{11}, b_{11} > 0$. Multiply the first row of $A \circ B$ by $\frac{a_{21}b_{21}}{a_{11}b_{11}}$ and subtract it from the second row to obtain

$$D = \begin{pmatrix} a_{11}b_{11} & a_{12}b_{12} & 0 & \cdots & 0 \\ 0 & a_{22}b_{22} - \frac{a_{12}b_{12}a_{21}b_{21}}{a_{11}b_{11}} & a_{23}b_{23} & \cdots & 0 \\ 0 & a_{32}b_{32} & a_{33}b_{33} & \cdots & 0 \\ \vdots & \vdots & \vdots & \ddots & \vdots \\ 0 & 0 & 0 & \cdots & a_{nn}b_{nn} \end{pmatrix}.$$

Therefore

$$(A \circ B)\begin{pmatrix} 1,\ldots,n \\ 1,\ldots,n \end{pmatrix} = a_{11}b_{11}D\begin{pmatrix} 2,\ldots,n \\ 2,\ldots,n \end{pmatrix}.$$

We now apply the induction hypothesis to the $(n-1)\times(n-1)$ submatrix of D obtained by deleting its first row and column. This matrix is the

Hadamard product of the two $(n-1)\times(n-1)$ totally positive Jacobi matrices

$$A_{b_{11}b_{22},b_{12}b_{21}}\begin{bmatrix}2,\ldots,n\\2,\ldots,n\end{bmatrix} \quad \text{and} \quad B\begin{bmatrix}2,\ldots,n\\2,\ldots,n\end{bmatrix}.$$

The former matrix is totally positive by our previous analysis and since $b_{11}b_{22}\geq b_{12}b_{21}$. This proves the proposition. □

Remark The same proof shows that the Hadamard product of a totally positive Jacobi matrix and an arbitrary totally positive matrix is a totally positive Jacobi matrix.

4. Hankel matrices. We are given Hankel matrices $A=(a_{i+j})_{i,j=0}^n$ and $B=(b_{i+j})_{i,j=0}^n$. If A and B are strictly totally positive, then $A\circ B$ is strictly totally positive. This follows from the fact that

$$A\circ B=(a_{i+j}b_{i+j})_{i,j=0}^n$$

is also a Hankel matrix, the characterization of strictly totally positive Hankel matrices as detailed in Theorem 4.4 and the Schur Product Theorem which states that if C and D are two strictly positive definite matrices, then so is their Hadamard product $C\circ D$.

5. Toeplitz matrices. In Theorem 4.5 we considered Toeplitz matrices of the form

$$A=(a_{j-i})_{i,j=1}^{\infty}$$

where $a_0>0$, $a_n\neq 0$ and $a_k=0$ for $k<0$ and $k>n$, and showed that A is totally positive if and only if

$$p(x)=\sum_{k=0}^{n}a_kx^k$$

has n negative zeros. A special case of Maló's Theorem states that if

$$p(x)=\sum_{k=0}^{n}a_kx^k$$

has n negative zeros and

$$q(x)=\sum_{k=0}^{m}b_kx^k$$

has m negative zeros, then the polynomial

$$\sum_{k=0}^{r} a_k b_k x^k$$

has r negative zeros, where $r = \min\{n, m\}$. Thus the Hadamard product of two totally positive Toeplitz matrices of the above form is again a totally positive Toeplitz matrix.

6. Hurwitz matrices. As stated in the remark at the end of Section 4.8, the polynomial

$$p(x) = \sum_{k=0}^{n} a_k x^k$$

is a Hurwitz polynomial if and only if all its zeros lie in the open left-hand plane $\mathrm{Re}\,(z) < 0$. Assuming $a_0 > 0$, we have that p is a Hurwitz polynomial if and only if the matrix

$$P = \begin{pmatrix} a_1 & a_3 & a_5 & \cdots \\ a_0 & a_2 & a_4 & \cdots \\ 0 & a_1 & a_3 & \cdots \\ 0 & a_0 & a_2 & \cdots \\ 0 & 0 & a_1 & \cdots \\ \vdots & \vdots & \vdots & \ddots \end{pmatrix}$$

satisfies

$$P\begin{pmatrix} 1, \ldots, r \\ 1, \ldots, r \end{pmatrix} > 0, \qquad r = 1, \ldots, n,$$

which implies that P is also totally positive (see Theorem 4.6). It is known that if

$$p(x) = \sum_{k=0}^{n} a_k x^k$$

is a Hurwitz polynomial and

$$q(x) = \sum_{k=0}^{m} b_k x^k$$

is a Hurwitz polynomial, then the polynomial

$$\sum_{k=0}^{r} a_k b_k x^k$$

is also a Hurwitz polynomial, where $r = \min\{n, m\}$. Thus the Hadamard product of two totally positive Hurwitz matrices of the above form is again a totally positive Hurwitz matrix of the appropriate form.

4.11 Remarks

The main reference books on totally positive kernels and Chebyshev systems are Gantmacher, Krein [1950], Karlin [1968], Karlin, Studden [1966] and Krein, Nudel'man [1977]. The examples of Sections 4.2 through 4.6 can all be found, in more or less detail, in Gantmacher, Krein [1937]. The examples of Sections 4.2 and 4.3 can also be found in Pólya, Szegő [1976] (originally published in 1925). Karlin [1968] contains many, many additional examples of totally positive and strictly totally positive kernels and matrices. See also Carlson, Gustafson [1983] for some further examples.

For a discussion of the theory of moments as presented in the proof of Theorem 4.4, see Karlin, Studden [1968], Chap. V, Krein, Nudel'man [1977], Chap. V and Shohat, Tamarkin [1943]. The form of the representation of the generating functions for totally positive infinite Toeplitz matrices was conjectured by Schoenberg in 1951. It was partially solved in Aissen, Schoenberg, Whitney [1952]. The final proof of the representations is to be found in Edrei [1952] and Edrei [1953]. The sequences in the totally positive Toeplitz matrices are also called *Pólya Frequency Sequences*; see Karlin [1968], Chap. 8, and references therein. For many more examples of sequences satisfying Theorem 4.5, see e.g., Brenti [1989], Brenti [1995], Pitman [1997], and Wang, Yeh [2005]. Theorem 4.6 is from Goodman, Sun [2004]. The fact that in the special case $M = 2$ (Hurwitz polynomials) the initial minors being strictly positive implies the total positivity of the full matrix was first proved by Asner [1970] and then reproved in a more transparent form by Kemperman [1982]. An example of a polynomial p whose zeros are not all in the closed left-hand plane $\mathrm{Re}\,(z) \leq 0$, but where the associated Hurwitz matrix P is totally positive (but without the strict positivity of the appropriate principal minors), can be found in Asner [1970]. The proof of Theorem 4.6 follows the lines of Kemperman's proof. See also Holtz [2003] and references therein for another approach to the Hurwitz polynomials and the total positivity of the associated matrix. Material on Hurwitz matrices may be found in various texts, e.g., Gantmacher [1953], Marden [1966], and Rahman, Schmeisser [2002]. Theorem 4.8 is a special case of a more general result; see e.g., Gantmacher [1953], Theorem 13, Chap. XV of the English translation.

Theorem 4.10 and Proposition 4.11 can be found, for example, in Rahman, Schmeisser [2002], Cor. 10.6.13 (the notation is somewhat different).

Applying the criteria of Theorem 2.16 we obtain relatively simple sufficient conditions for when certain of the matrices considered in this chapter are totally positive. For example, the Hankel matrix

$$A = (b_{i+j})_{i,j=0}^{n}$$

is strictly totally positive if $b_k > 0$, $k = 0, 1, \ldots$, and

$$b_{k-1}b_{k+1} > \left(4\cos^2 \frac{\pi}{n+2}\right) b_k^2, \qquad k = 1, \ldots, 2n-1,$$

while the Toeplitz matrix

$$A = (a_{j-i})_{i,j=1}^{\infty}$$

with $a_k = 0$ for $k < 0$ and $k > n$ is totally positive if $a_k > 0$, $k = 0, \ldots, n$, and

$$a_k^2 \geq 4a_{k-1}a_{k+1}, \qquad k = 1, \ldots, n-1.$$

Thus this condition is sufficient to imply that the polynomial

$$p(x) = \sum_{k=0}^{n} a_k x^k$$

has all real negative zeros. This was already proved in Kurtz [1992].

For a general discussion and survey of Hadamard products see Horn, Johnson [1991], Chap. 5, and the many references therein. It seems that Markham [1970] was the first to consider Hadamard products of totally positive matrices. He showed that the class of totally positive matrices is not closed under the operation of Hadamard products and proved Proposition 4.12. The Schur Product Theorem concerning the Hadamard product of positive definite matrices can be found, for example, in Horn, Johnson [1991], p. 309. Maló's Theorem is from Maló [1895]. In Wagner [1992] the result concerning the Hadamard product of Toeplitz matrices is extended to a subclass of the totally positive Toeplitz matrices with an infinite number of nonzero coefficients. The closure of Hurwitz matrices under the Hadamard product is in Garloff, Wagner [1996a]. Its proof is far beyond the scope of this monograph. Some additional results can be found in Garloff, Wagner [1996b] and Crans, Fallat, Johnson [2001].

5
Eigenvalues and eigenvectors

In this chapter we review the spectral properties of totally positive matrices. A strictly totally positive matrix has positive, simple eigenvalues and the associated eigenvectors possess an intricate structure. Such is not the case for totally positive matrices. However, there is an intermediate set of matrices with the same spectral properties as strictly totally positive matrices. These matrices are called oscillation matrices. They shall be discussed in Section 5.1. In Section 5.2 we present the Gantmacher–Krein Theorem (Theorem 5.3) and give two quite different proofs thereof. This theorem contains the main spectral properties of oscillation matrices. In Section 5.3 we consider eigenvalues of the principal submatrices of such matrices and study their behaviour. We study in more detail the properties of eigenvectors of oscillation matrices in Section 5.4. Finally, in Section 5.5, we look at how the eigenvalues of oscillation matrices vary as functions of the elements of the matrix.

5.1 Oscillation matrices

Oscillation matrices are a class of matrices intermediary between totally positive and strictly totally positive matrices. They share the eigenvalue and eigenvector structure of strictly totally positive matrices.

Definition 5.1 An $n \times n$ matrix A is said to be an *oscillation matrix* if A is totally positive and some power of A is strictly totally positive.

Importantly, there are relatively simple criteria for determining if a totally positive matrix is an oscillation matrix.

Theorem 5.2 *An $n \times n$ matrix $A = (a_{ij})_{i,j=1}^n$ is an oscillation matrix if and only if A is totally positive, nonsingular, and $a_{i,i+1}, a_{i+1,i} > 0$, $i = 1, \ldots, n-1$. Furthermore, if A is an oscillation matrix, then A^{n-1} is strictly totally positive.*

Proof The necessity of the above conditions is rather simple. If A is an oscillation matrix then A is totally positive. If A is singular then every power of A is singular and thus no power of A can be strictly totally positive. Assume A is totally positive and nonsingular and $a_{i,i+1} = 0$ for some i. Then from Theorem 1.19 we have $a_{rs} = 0$ for all $r \leq i$ and $s \geq i+1$. As is readily verified, this implies that the (r, s) elements of every power of A also vanish for all $r \leq i$ and $s \geq i+1$. This proves the necessity.

The sufficiency is less elementary. We first prove that

$$A\begin{pmatrix} i_1, \ldots, i_r \\ j_1, \ldots, j_r \end{pmatrix} > 0$$

for all $1 \leq i_1 < \cdots < i_r \leq n$ and $1 \leq j_1 < \cdots < j_r \leq n$ satisfying $|i_k - j_k| \leq 1$, $k = 1, \ldots, r$, and $\max\{i_k, j_k\} < \min\{i_{k+1}, j_{k+1}\}$, $k = 1, \ldots, r-1$. In this proof we make use of the fact that since A is totally positive and nonsingular, every principal minor of A is strictly positive (see Theorem 1.13). Our proof of this claim is via induction on r. The result holds for $r = 1$ since all of the elements of A on the main and first off-diagonals are strictly positive. Assume the result holds for $r-1$, but

$$A\begin{pmatrix} i_1, \ldots, i_r \\ j_1, \ldots, j_r \end{pmatrix} = 0$$

for some $i_1, \ldots, i_r$ and $j_1, \ldots, j_r$, as above. Then we must have $i_k \neq j_k$ for some k. From the induction hypothesis we have

$$A\begin{pmatrix} i_1, \ldots, i_{r-1} \\ j_1, \ldots, j_{r-1} \end{pmatrix}, \; A\begin{pmatrix} i_2, \ldots, i_r \\ j_2, \ldots, j_r \end{pmatrix} > 0.$$

Thus, from Proposition 1.17, the totally positive matrix

$$B = A\begin{bmatrix} i_1, i_1+1, \ldots, i_r \\ j_1, j_1+1, \ldots, j_r \end{bmatrix}$$

is of rank $r-1$. However, since $i_k \neq j_k$ for some k, and from the conditions on the $i_1, \ldots, i_r$ and $j_1, \ldots, j_r$, it follows that B contains an $r \times r$ singular principal submatrix of A. This is a contradiction since all principal minors of A are strictly positive.

From the above it easily follows that some power of A is strictly totally positive. However we wish to prove a bit more; namely that A^{n-1} is strictly totally positive. From Proposition 2.5 it suffices to prove that

$$A^{n-1}\begin{pmatrix} 1,\dots,r \\ n-r+1,\dots,n \end{pmatrix} > 0$$

and

$$A^{n-1}\begin{pmatrix} n-r+1,\dots,n \\ 1,\dots,r \end{pmatrix} > 0$$

for $r = 1,\dots,n$. We will only consider the former inequalities. Note that

$$A^{n-1}\begin{pmatrix} i_1,\dots,i_r \\ j_1,\dots,j_r \end{pmatrix} = \sum A\begin{pmatrix} i_1,\dots,i_r \\ s_1^1,\dots,s_r^1 \end{pmatrix} A\begin{pmatrix} s_1^1,\dots,s_r^1 \\ s_1^2,\dots,s_r^2 \end{pmatrix} \cdots A\begin{pmatrix} s_1^{n-2},\dots,s_r^{n-2} \\ j_1,\dots,j_r \end{pmatrix}$$

where the sum is over all $1 \le s_1^\ell < \cdots < s_r^\ell \le n$, $\ell = 1,\dots,n-2$. As each of these terms is nonnegative, for strict positivity to hold it suffices to prove that at least one of the products is strictly positive. For example,

$$A^{n-1}\begin{pmatrix} 1 \\ n \end{pmatrix} \ge A\begin{pmatrix} 1 \\ 2 \end{pmatrix} A\begin{pmatrix} 2 \\ 3 \end{pmatrix} \cdots A\begin{pmatrix} n-1 \\ n \end{pmatrix} > 0,$$

while

$$A^{n-1}\begin{pmatrix} 1,2 \\ n-1,n \end{pmatrix} \ge A\begin{pmatrix} 1,2 \\ 1,3 \end{pmatrix} A\begin{pmatrix} 1,3 \\ 2,4 \end{pmatrix} \cdots A\begin{pmatrix} n-3,n-1 \\ n-2,n \end{pmatrix} A\begin{pmatrix} n-2,n \\ n-1,n \end{pmatrix} > 0.$$

The strict positivity is a consequence of what we proved above.

Set

$$\mathbf{s}^\ell = (s_1^\ell,\dots,s_r^\ell),$$

where

$$s_k^\ell = \min\{n-r+k, k+(\ell+k-r)_+\}, \qquad k = 1,\dots,r, \quad \ell = 0,\dots,n-1.$$

Here $m_+ = \max\{m,0\}$. It is now readily verified that $|s_k^\ell - s_k^{\ell+1}| \le 1$ and

$$\max\{s_k^\ell, s_k^{\ell+1}\} = s_k^{\ell+1} < s_{k+1}^\ell = \min\{s_{k+1}^\ell, s_{k+1}^{\ell+1}\}.$$

Thus

$$A\begin{pmatrix} s_1^\ell,\dots,s_r^\ell \\ s_1^{\ell+1},\dots,s_r^{\ell+1} \end{pmatrix} > 0.$$

As $\mathbf{s}^0 = (1, \ldots, r)$ and $\mathbf{s}^{n-1} = (n-r+1, \ldots, n)$ it follows that

$$A^{n-1}\begin{pmatrix} 1, \ldots, r \\ n-r+1, \ldots, n \end{pmatrix} > 0. \qquad \square$$

Remark If A_k, $k = 1, \ldots, n-1$, are arbitrary $n \times n$ oscillation matrices, then we also have that $A_1 \cdots A_{n-1}$ is strictly totally positive. The same proof applies.

5.2 The Gantmacher–Krein theorem

The main theorem concerning spectral properties of oscillation matrices is the following.

Theorem 5.3 (Gantmacher–Krein). *Let A be an $n \times n$ oscillation matrix. Then the n eigenvalues of A are positive and simple. In addition, if we denote by $\mathbf{u}^k$ a real eigenvector (unique up to multiplication by a nonzero constant) associated with the eigenvalue λ_k, where $\lambda_1 > \lambda_2 > \cdots > \lambda_n > 0$, then*

$$q - 1 \le S^-\left(\sum_{i=q}^{p} c_i \mathbf{u}^i\right) \le S^+\left(\sum_{i=q}^{p} c_i \mathbf{u}^i\right) \le p - 1,$$

for each $1 \le q \le p \le n$ (and c_i not all zero). In particular, $S^-(\mathbf{u}^k) = S^+(\mathbf{u}^k) = k - 1$ for $k = 1, \ldots, n$.

For the definitions of S^- and S^+, see Chapter 3.

We present two very different proofs of Theorem 5.3. The first is the classic proof. In this classic proof we use two known general results. The first of these is Perron's Theorem.

Perron's Theorem *Let A be an $n \times n$ matrix, all of whose elements are strictly positive. Then A has a simple, positive eigenvalue that is strictly greater in modulus than all other eigenvalues of A. Furthermore the unique (up to multiplication by a nonzero constant) associated eigenvector may be chosen so that all its components are strictly positive.*

Proof There are numerous proofs of this result in the literature, and an almost uncountable number of generalizations. For completeness, we present a proof that seems to be one of the simpler and more transparent.

For vectors $\mathbf{x}$ and $\mathbf{y}$ in $\mathbb{R}^n$, we write $\mathbf{x} \ge \mathbf{y}$ if $x_i \ge y_i$, $i = 1, \ldots, n$, and

$\mathbf{x} > \mathbf{y}$ if $x_i > y_i$, $i = 1, \ldots, n$. Set

$$\lambda^* = \sup\{\lambda \,:\, A\mathbf{x} \geq \lambda\mathbf{x}\,,\ \text{for some } \mathbf{x} \geq \mathbf{0}\}\,.$$

Since all elements of A are strictly positive, we must have $\lambda^* > 0$. A convergence (compactness) argument implies the existence of an $\mathbf{x}^* \geq \mathbf{0}$ ($\mathbf{x}^* \neq \mathbf{0}$) such that $A\mathbf{x}^* \geq \lambda^*\mathbf{x}^*$. If $A\mathbf{x}^* \neq \lambda^*\mathbf{x}^*$, then $A(A\mathbf{x}^*) > \lambda^*(A\mathbf{x}^*)$ since all entries of A are strictly positive. Setting $\mathbf{y}^* = A\mathbf{x}^*$ it follows that $A\mathbf{y}^* > \lambda^*\mathbf{y}^*$, which contradicts our definition of λ^*. Thus

$$A\mathbf{x}^* = \lambda^*\mathbf{x}^*\,.$$

Now $\mathbf{x}^* \geq \mathbf{0}$ ($\mathbf{x}^* \neq \mathbf{0}$) and thus $A\mathbf{x}^* > \mathbf{0}$, whence $\mathbf{x}^* > \mathbf{0}$. We have found a positive eigenvalue with a strictly positive eigenvector.

Let λ be any other eigenvalue of A. Then

$$A\mathbf{y} = \lambda\mathbf{y}$$

for some $\mathbf{y} \in \mathbb{C}^n \backslash \{\mathbf{0}\}$. Now

$$|\lambda|\,|\mathbf{y}| = |\lambda\mathbf{y}| = |A\mathbf{y}| \leq A|\mathbf{y}|\,,$$

where $|\mathbf{y}| = (|y_1|, \ldots, |y_n|)$. From the definition of λ^*, it follows that $|\lambda| \leq \lambda^*$. If $|\lambda| = \lambda^*$, then we must have $\lambda^*|\mathbf{y}| = A|\mathbf{y}|$ (since otherwise $\lambda^*(A|\mathbf{y}|) < A(A|\mathbf{y}|)$, contradicting the definition of λ^*). Thus, for each $i = 1, \ldots, n$,

$$|\lambda|\,|y_i| = \left|\sum_{j=1}^n a_{ij}y_j\right| = \sum_{j=1}^n a_{ij}|y_j|\,.$$

This implies the existence of a $\gamma \in \mathbb{C}$, $|\gamma| = 1$, such that $\gamma y_j = |y_j|$ for each $j = 1, \ldots, n$. Thus, we may in fact assume that if $|\lambda| = \lambda^*$, then $\lambda = |\lambda| = \lambda^*$, and for the associated eigenvector $\mathbf{y}$, we have $\mathbf{y} \geq \mathbf{0}$. Two consequences of these facts are the following: for every eigenvalue λ, $\lambda \neq \lambda^*$, we have $|\lambda| < \lambda^*$; the geometric multiplicity of the eigenvalue λ^* is exactly 1. The latter holds because, if not, we can easily construct a real eigenvector associated with λ^* that is not of one sign, and this contradicts the above analysis.

It remains to prove that the eigenvalue λ^* is of algebraic multiplicity 1. Assume not. There then exists a vector $\mathbf{y}^*$ (linearly independent of $\mathbf{x}^*$) such that

$$A\mathbf{y}^* = \lambda^*\mathbf{y}^* + \alpha\mathbf{x}^*$$

with some $\alpha \neq 0$. This $\mathbf{y}^*$ is called a "generalized eigenvector." The transpose of A, namely A^T, has the same eigenvalues as A and is obviously strictly positive. As such there exists an eigenvector $\mathbf{w}^* > \mathbf{0}$ associated with the eigenvalue λ^*. Now

$$\lambda^*(\mathbf{w}^*, \mathbf{y}^*) = (A^T\mathbf{w}^*, \mathbf{y}^*) = (\mathbf{w}^*, A\mathbf{y}^*) = \lambda^*(\mathbf{w}^*, \mathbf{y}^*) + \alpha(\mathbf{w}^*, \mathbf{x}^*).$$

Since $\mathbf{w}^*, \mathbf{x}^* > \mathbf{0}$ and $\alpha \neq 0$ we have $\alpha(\mathbf{w}^*, \mathbf{x}^*) \neq 0$, a contradiction. Thus the algebraic multiplicity of λ^* is one. □

The second general result we use in the proof of Theorem 5.3 is called Kronecker's Theorem. We recall from Chapter 1 that the pth *compound matrix*, denoted by $A_{[p]}$, of the $n \times n$ matrix A is defined as the $\binom{n}{p} \times \binom{n}{p}$ matrix with entries

$$(A(\mathbf{i}, \mathbf{j}))_{\mathbf{i} \in I_p^n, \mathbf{j} \in I_p^n}$$

where the $\mathbf{i} \in I_p^n$ and $\mathbf{j} \in I_p^n$ are arranged in lexicographic order, i.e., for $\mathbf{i}, \mathbf{j} \in I_p^n$ we set $\mathbf{i} \geq \mathbf{j}$ ($\mathbf{i} \neq \mathbf{j}$) if the first nonzero term in the sequence $i_1 - j_1, \ldots, i_p - j_p$ is positive.

Kronecker's Theorem *Let A be an $n \times n$ matrix with eigenvalues $\lambda_1, \ldots, \lambda_n$ listed to their algebraic multiplicity. Then the $\binom{n}{p}$ eigenvalues of $A_{[p]}$, listed to their algebraic multiplicity, are $\lambda_{i_1} \cdots \lambda_{i_p}$ for $1 \leq i_1 < \cdots < i_p \leq n$.*

Proof We present two proofs. Every $n \times n$ matrix A may be written in the form

$$A = P^{-1}TP$$

where T is an upper triangular matrix. As such, the diagonal entries of T are simply the eigenvalues of A. Now, from the Cauchy–Binet formula,

$$A_{[p]} = (P^{-1})_{[p]}T_{[p]}P_{[p]} = (P_{[p]})^{-1}T_{[p]}P_{[p]}.$$

The matrix $T_{[p]}$ is upper triangular and its diagonal entries are exactly the products of p distinct diagonal entries of T. This proves the result.

A second proof is the following. Associated with each eigenvalue λ_i is an eigenvector/generalized eigenvector $\mathbf{u}^i$, $i = 1, \ldots, n$, such that the $\mathbf{u}^1, \ldots, \mathbf{u}^n$ span $\mathbb{C}^n$. Now (see Chapter 1)

$$A_{[p]}\left(\mathbf{u}^{i_1} \wedge \cdots \wedge \mathbf{u}^{i_p}\right) = A\mathbf{u}^{i_1} \wedge \cdots \wedge A\mathbf{u}^{i_p}$$

from which it follows that the $\mathbf{u}^{i_1} \wedge \cdots \wedge \mathbf{u}^{i_p}$ are eigenvectors/generalized eigenvectors of $A_{[p]}$ with associated eigenvalues $\lambda_{i_1} \cdots \lambda_{i_p}$. Since these vectors are linearly independent, we have determined all the $\binom{n}{p}$ eigenvalues (and generalized eigenvectors) of $A_{[p]}$. □

We now present our first proof of Theorem 5.3.

Proof of Theorem 5.3 It suffices to prove the theorem for strictly totally positive matrices. For if A has eigenvalues $\lambda_1, \ldots, \lambda_n$, listed to their algebraic multiplicity, then A^k has the eigenvalues $\lambda_1^k, \ldots, \lambda_n^k$. If we show that $\lambda_1^k > \cdots > \lambda_n^k > 0$ for all k sufficiently large, then obviously we must have $\lambda_1 > \cdots > \lambda_n > 0$. In addition, if $A\mathbf{u} = \lambda\mathbf{u}$ then $A^k\mathbf{u} = \lambda^k\mathbf{u}$, so that if A has n distinct eigenvalues, then the eigenvectors of A and A^k are one and the same.

Let $\lambda_1, \ldots, \lambda_n$ denote the eigenvalues of A listed to their algebraic multiplicity, and assume that $|\lambda_1| \geq \cdots \geq |\lambda_n| \geq 0$. As A is strictly totally positive we have that $A_{[p]}$ is a matrix all of whose entries are strictly positive. From the Perron and Kronecker Theorems applied to $A_{[p]}$ we have

$$\lambda_1 \cdots \lambda_p > |\lambda_1 \cdots \lambda_{p-1}\lambda_{p+1}|$$

for all $p = 1, \ldots, n$. In addition, A is nonsingular and thus $|\lambda_n| > 0$. From these two facts we have

$$\lambda_1 > \cdots > \lambda_n > 0\,.$$

From the Perron and Kronecker Theorems it also follows that by a suitable normalization of the associated eigenvectors $\mathbf{u}^1, \ldots, \mathbf{u}^n$, we may assume that

$$\mathbf{u}^1 \wedge \cdots \wedge \mathbf{u}^p > \mathbf{0}$$

for $p = 1, 2, \ldots, n$.

It remains for us to prove the sign change properties of the eigenvectors $\mathbf{u}^1, \ldots, \mathbf{u}^n$. Assume $S^+\left(\sum_{i=q}^p c_i\mathbf{u}^i\right) \geq p$ for some choice of nontrivial $(c_q, \ldots, c_p)$. There then exist $j_0 < \cdots < j_p$ and an $\varepsilon \in \{-1, 1\}$ such that

$$\varepsilon(-1)^k \left(\sum_{i=q}^p c_i\mathbf{u}^i\right)_{j_k} \geq 0\,, \qquad k = 0, \ldots, p\,.$$

Set $\mathbf{u}^0 = \sum_{i=q}^p c_i\mathbf{u}^i$. Let $\mathbf{u}^i = (u_{1,i}, \ldots, u_{n,i})$, $i = 0, 1, \ldots, n$, and U denote the matrix with columns $\mathbf{u}^0, \mathbf{u}^1, \ldots, \mathbf{u}^n$. Thus

$$U\begin{pmatrix} j_0, j_1, \ldots, j_p \\ 0, 1, \ldots, p \end{pmatrix} = \det\,(u_{j_k,i})_{i,k=0}^p = 0$$

since $\mathbf{u}^0$ is a linear combination of the $\mathbf{u}^1, \ldots, \mathbf{u}^p$. On the other hand, when expanding this matrix by the first column we obtain

$$U\begin{pmatrix} j_0, j_1, \ldots, j_p \\ 0, 1, \ldots, p \end{pmatrix} = \sum_{k=0}^{p} (-1)^k u_{j_k,0} U \begin{pmatrix} j_0, \ldots, \widehat{j_k}, \ldots, j_p \\ 1, \ldots, p \end{pmatrix}.$$

Since $\mathbf{u}^1 \wedge \cdots \wedge \mathbf{u}^p > \mathbf{0}$, we have

$$U\begin{pmatrix} j_0, \ldots, \widehat{j_k}, \ldots, j_p \\ 1, \ldots, p \end{pmatrix} > 0,$$

for each $k = 0, \ldots, p$. As $(-1)^k u_{j_k,0}$ is weakly of one sign, we must have $u_{j_k,0} = 0$, for all $k = 0, \ldots, p$. But this is impossible since $\mathbf{u}^1 \wedge \cdots \wedge \mathbf{u}^p > \mathbf{0}$ implies that all $p \times p$ minors of the $n \times p$ matrix with columns $\mathbf{u}^1, \ldots, \mathbf{u}^p$ are nonsingular. In other words, no nontrivial linear combination of the $\mathbf{u}^1, \ldots, \mathbf{u}^n$ vanishes at p coordinates. Thus

$$S^+\left(\sum_{i=q}^{p} c_i \mathbf{u}^i\right) \le p - 1.$$

Let $\mathbf{v}^1, \ldots, \mathbf{v}^n$ be left eigenvectors of A with associated eigenvalues $\lambda_1, \ldots, \lambda_n$, respectively. We assume, by what we have proved so far, that we have normalized the $\{\mathbf{v}^j\}_{j=1}^n$ so that $\mathbf{v}^1 \wedge \cdots \wedge \mathbf{v}^p > \mathbf{0}$ for $p = 1, \ldots, n$. Thus, in particular,

$$S^+\left(\sum_{j=1}^{p} b_j \mathbf{v}^j\right) \le p - 1$$

for every choice of nontrivial $(b_1, \ldots, b_p)$, $p = 1, \ldots, n$.

Let $\mathbf{u} \in \mathbb{R}^n \backslash \{\mathbf{0}\}$ be any vector satisfying

$$(\mathbf{u}, \mathbf{v}^j) = 0, \qquad j = 1, \ldots, q-1.$$

We claim that $S^-(\mathbf{u}) \ge q-1$. If $S^-(\mathbf{u}) = r \le q-2$, then there exist indices $1 \le i_1 < \cdots < i_r < n$ and an $\varepsilon \in \{-1, 1\}$ such that

$$\varepsilon(-1)^k u_j \ge 0, \qquad i_{k-1} + 1 \le j \le i_k, \qquad k = 1, \ldots, r+1,$$

and $u_j \ne 0$ for some $i_{k-1} + 1 \le j \le i_k$, and each $k = 1, \ldots, r+1$. Here $i_0 = 0$ and $i_{r+1} = n$. Let

$$\mathbf{v} = \sum_{j=1}^{r+1} b_j \mathbf{v}^j$$

satisfy $\mathbf{v} \neq \mathbf{0}$ and $v_{i_k} = 0$, $k = 1, \ldots, r$. Such a $\mathbf{v}$ exists. Since $S^+\left(\sum_{j=1}^{r+1} b_j \mathbf{v}^j\right) \leq r$, we must have

$$\delta(-1)^k v_j > 0\,, \qquad i_{k-1}+1 \leq j < i_k, \qquad k = 1, \ldots, r+1,$$

and also $\delta(-1)^{r+1} v_n > 0$, where $\delta \in \{-1, 1\}$. As $r+1 \leq q-1$ and

$$(\mathbf{u}, \mathbf{v}) \neq 0\,,$$

this is a contradiction, implying that $S^-(\mathbf{u}) \geq q-1$.

The calculation

$$\lambda_i(\mathbf{u}^i, \mathbf{v}^j) = (\lambda_i \mathbf{u}^i, \mathbf{v}^j) = (A\mathbf{u}^i, \mathbf{v}^j) = (\mathbf{u}^i, \mathbf{v}^j A) = \lambda_j(\mathbf{u}^i, \mathbf{v}^j),$$

implies that $(\mathbf{u}^i, \mathbf{v}^j) = 0$ for $i \neq j$. Thus

$$\left(\sum_{i=q}^{p} c_i \mathbf{u}^i, \mathbf{v}^j\right) = 0\,, \qquad j = 1, \ldots, q-1\,,$$

and therefore

$$S^-\left(\sum_{i=q}^{p} c_i \mathbf{u}^i\right) \geq q-1\,,$$

which proves the theorem.

Here is an alternative proof of this lower bound. The matrix A^{-1} is not a strictly totally positive matrix. However, $C = DA^{-1}D$ is strictly totally positive where D is the diagonal matrix with diagonal entries alternately 1 and -1 (see Proposition 1.6). For $\mathbf{u} = (u_1, \ldots, u_n)$, set $\widetilde{\mathbf{u}} = D\mathbf{u} = (u_1, -u_2, \ldots, (-1)^{n+1}u_n)$, i.e., component u_j is replaced by $(-1)^{j-1}u_j$. From Lemma 3.1

$$S^-(\mathbf{u}) + S^+(\widetilde{\mathbf{u}}) = n-1$$

for every $\mathbf{u} \in \mathbb{R}^n$, $\mathbf{u} \neq \mathbf{0}$. As $\mathbf{u}^k$ is an eigenvector of A with associated eigenvalue λ_k, the strictly totally positive matrix C has the eigenvector $\widetilde{\mathbf{u}}^k$ with associated eigenvalue $1/\lambda_k$. Because

$$\frac{1}{\lambda_n} > \cdots > \frac{1}{\lambda_1} > 0$$

it follows from our previous result that

$$S^+\left(\sum_{i=q}^{p} c_i \widetilde{\mathbf{u}}^i\right) \leq n-q,$$

implying

$$S^-\left(\sum_{i=q}^{p} c_i\mathbf{u}^i\right) \geq q-1\,. \qquad \square$$

We now present a very different proof of Theorem 5.3. The proof of the sign change properties of the eigenvectors will be based on the variation diminishing properties of strictly totally positive matrices. We start by providing an alternative proof of the fact that A has n simple and positive eigenvalues. We also simultaneously prove that the eigenvalues of the two principal submatrices obtained by deleting from A either the first row and column, or the last row and column, strictly interlace the eigenvalues of A.

Proposition 5.4 *Let A be an $n \times n$ strictly totally positive matrix. Then its n eigenvalues are positive and simple. In addition, if these eigenvalues are denoted by $\lambda_1 > \lambda_2 > \cdots > \lambda_n > 0$, and $\mu_1^{(k)} > \cdots > \mu_{n-1}^{(k)} > 0$ are the eigenvalues of the principal submatrix of A obtained by deleting its kth row and column, then*

$$\lambda_j > \mu_j^{(k)} > \lambda_{j+1}\,, \qquad j = 1, \ldots, n-1\,,$$

for $k = 1$ and $k = n$.

Proof For ease of notation let $A_\lambda = A - \lambda I$. From Sylvester's Determinant Identity we have

$$\begin{aligned} &A_\lambda\begin{pmatrix}1,\ldots,n\\1,\ldots,n\end{pmatrix} A_\lambda\begin{pmatrix}2,\ldots,n-1\\2,\ldots,n-1\end{pmatrix} \\ &= A_\lambda\begin{pmatrix}1,\ldots,n-1\\1,\ldots,n-1\end{pmatrix} A_\lambda\begin{pmatrix}2,\ldots,n\\2,\ldots,n\end{pmatrix} \\ &\quad - A_\lambda\begin{pmatrix}1,\ldots,n-1\\2,\ldots,n\end{pmatrix} A_\lambda\begin{pmatrix}2,\ldots,n\\1,\ldots,n-1\end{pmatrix}. \end{aligned} \tag{5.1}$$

It is readily verified that

$$A_\lambda\begin{pmatrix}1,\ldots,n-1\\2,\ldots,n\end{pmatrix},\quad A_\lambda\begin{pmatrix}2,\ldots,n\\1,\ldots,n-1\end{pmatrix} > 0 \tag{5.2}$$

for all $\lambda > 0$. Simply expand each determinant as a polynomial in λ and note that all the coefficients of this polynomial are strictly positive since all minors of A are strictly positive.

We now use an induction argument to prove our result. The case $n = 2$

is easily checked by hand. As such we assume that $n > 2$. For notational ease, set

$$\begin{aligned} p(\lambda) &= A_\lambda \begin{pmatrix} 1,\ldots,n \\ 1,\ldots,n \end{pmatrix} \\ q_1(\lambda) &= A_\lambda \begin{pmatrix} 2,\ldots,n \\ 2,\ldots,n \end{pmatrix}, \\ q_2(\lambda) &= A_\lambda \begin{pmatrix} 1,\ldots,n-1 \\ 1,\ldots,n-1 \end{pmatrix} \\ r(\lambda) &= A_\lambda \begin{pmatrix} 2,\ldots,n-1 \\ 2,\ldots,n-1 \end{pmatrix}. \end{aligned}$$

Thus from (5.1) and (5.2)

$$p(\lambda)r(\lambda) < q_1(\lambda)q_2(\lambda) \tag{5.3}$$

for all $\lambda > 0$.

We assume, by the induction hypothesis, that the $n-2$ zeros of r are positive and simple, and interlace the $n-1$ positive simple zeros of both q_1 and q_2. Let

$$\mu_1 > \cdots > \mu_{n-1} > 0$$

denote the zeros of either q_1 or q_2, i.e., $\mu_j = \mu_j^{(1)}$ or $\mu_j = \mu_j^{(n)}$, $j = 1,\ldots,n-1$. From (5.3) it follows that

$$p(\mu_i)r(\mu_i) < 0\,, \qquad i = 1,\ldots,n-1\,.$$

Furthermore, as $r(0) > 0$ and by the induction hypothesis,

$$r(\mu_i)(-1)^{i+n-1} > 0\,, \qquad i = 1,\ldots,n-1\,.$$

We therefore have

$$p(\mu_i)(-1)^{i+n} > 0\,, \qquad i = 1,\ldots,n-1\,.$$

As $p(0) > 0$ and $p(\mu_{n-1}) < 0$, the polynomial p has an additional zero in $(0, \mu_{n-1})$. Furthermore, the polynomial p has leading coefficient $(-1)^n$ and $p(\mu_1)(-1)^{n+1} > 0$. Thus p has an additional zero in (μ_1, ∞). As such, p has n positive, simple zeros that are interlaced by the $\{\mu_j\}_{j=1}^{n-1}$. This advances the induction step. □

Another Proof of Theorem 5.3 We assume that $\lambda > \mu > 0$ are eigenvalues of A with associated eigenvectors $\mathbf{x}$ and $\mathbf{y}$. Then from Theorem 3.3 we have for all $(\alpha, \beta) \neq (0,0)$

$$S^+(\lambda\alpha\mathbf{x} + \mu\beta\mathbf{y}) = S^+(A(\alpha\mathbf{x} + \beta\mathbf{y})) \leq S^-(\alpha\mathbf{x} + \beta\mathbf{y})\,. \tag{5.4}$$

Note that this implies, taking $\beta = 0$ and $\alpha = 0$, respectively, that

$$S^+(\mathbf{x}) = S^-(\mathbf{x}) \text{ and } S^+(\mathbf{y}) = S^-(\mathbf{y}).$$

Assume that both $\alpha \neq 0$ and $\beta \neq 0$. We may rewrite (5.4) as

$$S^+\left(\alpha\mathbf{x} + \left(\frac{\mu}{\lambda}\right)\beta\mathbf{y}\right) \leq S^-(\alpha\mathbf{x} + \beta\mathbf{y}).$$

Iterating this process we have

$$S^+\left(\alpha\mathbf{x} + \left(\frac{\mu}{\lambda}\right)^k\beta\mathbf{y}\right) \leq S^-(\alpha\mathbf{x} + \beta\mathbf{y}),$$

for every positive integer k. As $k \to \infty$, and since $\mu/\lambda < 1$, we have from Lemma 3.2

$$\varliminf_{k\to\infty} S^+\left(\alpha\mathbf{x} + \left(\frac{\mu}{\lambda}\right)^k\beta\mathbf{y}\right) \geq S^-(\mathbf{x}).$$

(In fact, since $S^-(\mathbf{x}) = S^+(\mathbf{x})$ it follows that

$$\lim_{k\to\infty} S^+\left(\alpha\mathbf{x} + \left(\frac{\mu}{\lambda}\right)^k\beta\mathbf{y}\right) = S^-(\mathbf{x}).)$$

Similarly, from (5.4)

$$S^+(\alpha\mathbf{x} + \beta\mathbf{y}) \leq S^-\left(\left(\frac{\mu}{\lambda}\right)\alpha\mathbf{x} + \beta\mathbf{y}\right),$$

and iterating this process we obtain

$$S^+(\alpha\mathbf{x} + \beta\mathbf{y}) \leq S^-\left(\left(\frac{\mu}{\lambda}\right)^k\alpha\mathbf{x} + \beta\mathbf{y}\right),$$

for every positive integer k. As $k \to \infty$, and since $\mu/\lambda < 1$ and $S^-(\mathbf{y}) = S^+(\mathbf{y})$, it follows (see Lemma 3.2) that

$$\varlimsup_{k\to\infty} S^-\left(\left(\frac{\mu}{\lambda}\right)^k\alpha\mathbf{x} + \beta\mathbf{y}\right) \leq S^+(\mathbf{y}) = S^-(\mathbf{y}).$$

Thus for every $(\alpha, \beta) \neq (0, 0)$,

$$S^-(\mathbf{x}) \leq S^-(\alpha\mathbf{x} + \beta\mathbf{y}) \leq S^+(\alpha\mathbf{x} + \beta\mathbf{y}) \leq S^-(\mathbf{y}).$$

If $S^-(\mathbf{x}) = S^-(\mathbf{y})$, then equality holds throughout our sequence of inequalities, and in particular

$$S^+(\alpha\mathbf{x} + \beta\mathbf{y}) = S^-(\alpha\mathbf{x} + \beta\mathbf{y})$$

for all $(\alpha, \beta) \neq (0, 0)$. This is impossible since there are choices of α, β for which the first or last coefficient of the vector $\alpha\mathbf{x} + \beta\mathbf{y}$ vanishes, in which case the above equality cannot possibly hold. Therefore $S^-(\mathbf{x}) < S^-(\mathbf{y})$.

We proved in Proposition 5.4 that A has n simple, positive eigenvalues and thus n associated eigenvectors. For each of these n eigenvectors we have $S^+(\mathbf{x}) = S^-(\mathbf{x}) \in \{0, 1, \ldots, n-1\}$, and if $\lambda > \mu > 0$ are eigenvalues with associated eigenvectors $\mathbf{x}$ and $\mathbf{y}$, then $S^-(\mathbf{x}) < S^-(\mathbf{y})$. This proves that if

$$\lambda_1 > \cdots > \lambda_n > 0\,,$$

are the eigenvalues of A, and $\mathbf{u}^k$ is a real eigenvector associated with the eigenvalue λ_k, then necessarily

$$S^+(\mathbf{u}^k) = S^-(\mathbf{u}^k) = k-1, \qquad k = 1, \ldots, n.$$

Now assume that $1 \le q \le p \le n$ and we are given real constants $c_q, \ldots, c_p$, not all zero. Then, following the above analysis, repeated application of

$$S^+\left(\sum_{i=q}^{p} c_i \lambda_i \mathbf{u}^i\right) = S^+\left(\sum_{i=q}^{p} c_i A\mathbf{u}^i\right) \le S^-\left(\sum_{i=q}^{p} c_i \mathbf{u}^i\right)$$

implies that for every positive integer m

$$\begin{aligned} S^+\left(\sum_{i=q}^{p} c_i \left(\frac{\lambda_i}{\lambda_q}\right)^m \mathbf{u}^i\right) &\le S^-\left(\sum_{i=q}^{p} c_i \mathbf{u}^i\right) \le S^+\left(\sum_{i=q}^{p} c_i \mathbf{u}^i\right) \\ &\le S^-\left(\sum_{i=q}^{p} c_i \left(\frac{\lambda_p}{\lambda_i}\right)^m \mathbf{u}^i\right). \end{aligned}$$

We may assume that $c_p, c_q \neq 0$, and apply Lemma 3.2 to obtain

$$\varliminf_{m\to\infty} S^+\left(\sum_{i=q}^{p} c_i \left(\frac{\lambda_i}{\lambda_q}\right)^m \mathbf{u}^i\right) \ge S^-(\mathbf{u}^q) = q-1\,,$$

while

$$\varlimsup_{m\to\infty} S^-\left(\sum_{i=q}^{p} c_i \left(\frac{\lambda_p}{\lambda_i}\right)^m \mathbf{u}^i\right) \le S^+(\mathbf{u}^p) = p-1\,.$$

Thus

$$q-1 \le S^-\left(\sum_{i=q}^{p} c_i \mathbf{u}^i\right) \le S^+\left(\sum_{i=q}^{p} c_i \mathbf{u}^i\right) \le p-1\,,$$

which completes the proof. □

Remark There is one more result that is sometimes stated in connection with the Gantmacher–Krein Theorem (Theorem 5.3). It has to do with the "interlacing of the zeros" of $\mathbf{u}^{k+1}$ and $\mathbf{u}^k$, which is important in the continuous (integral) analogue of this theorem. One form of this is the following. Let $u_k(t)$ be the continuous function defined on $[1, n]$ which is linear on each $[j, j+1]$, $j = 1, \ldots, n-1$, and such that $u_k(j)$ is the jth component of $\mathbf{u}^k$. Since

$$S^-(\mathbf{u}^k) = S^+(\mathbf{u}^k) = k - 1\,,$$

the function $u_k(t)$ has exactly $k-1$ zeros that are each strict sign changes. Since

$$k - 1 \le S^-(\alpha\mathbf{u}^k + \beta\mathbf{u}^{k+1}) \le S^+(\alpha\mathbf{u}^k + \beta\mathbf{u}^{k+1}) \le k\,,$$

for every choice of $(\alpha, \beta) \neq (0, 0)$, it can be shown that the $k-1$ zeros of $u_k(t)$ strictly interlace the k zeros of $u_{k+1}(t)$.

Since strictly totally positive matrices are dense in the set of totally positive matrices, see Theorem 2.6, and because of the continuity of the eigenvalues as functions of the matrix entries, we have the following:

Corollary 5.5 *The eigenvalues of totally positive matrices are both real and nonnegative.*

5.3 Eigenvalues of principal submatrices

What can we say about the eigenvalues of the principal submatrices of a strictly totally positive matrix? Let

$$\lambda_1 > \cdots > \lambda_n > 0$$

denote the eigenvalues of A, and

$$\mu_1^{(k)} > \cdots > \mu_{n-1}^{(k)} > 0$$

the eigenvalues of the principal submatrix of A obtained by deleting its kth row and column. As we have seen in Proposition 5.4, the $n-1$ positive, simple eigenvalues $\{\mu_i^{(1)}\}_{i=1}^{n-1}$ and $\{\mu_i^{(n)}\}_{i=1}^{n-1}$ strictly interlace those of the original matrix A. This naturally begs the question of whether this same property holds for all principal submatrices? It also easily follows (see the proof of the Perron Theorem on positive matrices) that

$$\lambda_1 > \mu_1^{(k)}$$

for each k. However, the strict interlacing property does not hold for all principal submatrices. As an example, consider

$$A = \begin{pmatrix} 1 & 1 & 0 \\ 2 & 2 & 1 \\ 2 & 2 & 1 \end{pmatrix}.$$

The matrix A is totally positive but not strictly totally positive. However, since strictly totally positive matrices are dense in the class of totally positive matrices, there are strictly totally positive matrices whose eigenvalues (and the eigenvalues of its principal submatrices) are arbitrarily close to those of A. The eigenvalues of A are $2+\sqrt{3}, 2-\sqrt{3}$, and 0. The eigenvalues of the principal submatrix of A obtained by deleting the second row and column are 1 and 1. Obviously we do not have interlacing.

Nevertheless there is a weaker interlacing that does hold and it is the following.

Theorem 5.6 *Let A be an $n \times n$ strictly totally positive matrix. Then for each k, $1 < k < n$,*

$$\lambda_{j-1} > \mu_j^{(k)} > \lambda_{j+1}\,, \qquad j = 1, \ldots, n-1\,,$$

(where $\lambda_0 = \lambda_1$).

Proof We first prove that for any k and j as above,

$$\mu_j^{(k)} > \lambda_{j+1}.$$

Let A_k denote the principal submatrix of A obtained by deleting the kth row and column. Let $\mathbf{v}$ denote a real eigenvector of A_k with eigenvalue $\mu_j^{(k)}$, i.e.,

$$A_k\mathbf{v} = \mu_j^{(k)}\mathbf{v}\,.$$

From Theorem 5.3

$$S^+(\mathbf{v}) = S^-(\mathbf{v}) = j-1\,.$$

We construct the vector $\mathbf{v}' \in \mathbb{R}^n$ from the vector $\mathbf{v} \in \mathbb{R}^{n-1}$ by simply inserting a 0 as its new kth coordinate. That is, we write (somewhat abusing notation)

$$\mathbf{v} = (v_1, \ldots, v_{k-1}, v_{k+1}, \ldots, v_n)$$

and set

$$\mathbf{v}' = (v_1, \ldots, v_{k-1}, 0, v_{k+1}, \ldots, v_n)\,.$$

Let $\mathbf{v}''$ be defined by

$$A\mathbf{v}' = \mu_j^{(k)}\mathbf{v}''.$$

Thus

$$\mathbf{v}'' = (v_1, \ldots, v_{k-1}, w_k, v_{k+1}, \ldots, v_n),$$

for some easily calculated w_k. From Theorem 3.3 and the properties of $\mathbf{v}$, $\mathbf{v}'$ and $\mathbf{v}''$ we have,

$$j - 1 = S^+(\mathbf{v}) \le S^+(\mathbf{v}'') \le S^-(\mathbf{v}') = S^-(\mathbf{v}) = j - 1.$$

Thus

$$S^+(\mathbf{v}'') = S^-(\mathbf{v}') = j - 1.$$

Let $\mathbf{u}$ denote a real eigenvector of A with eigenvalue λ_{j+1}, i.e.,

$$A\mathbf{u} = \lambda_{j+1}\mathbf{u}.$$

Then from Theorem 5.3

$$S^+(\mathbf{u}) = S^-(\mathbf{u}) = j.$$

If $u_k = 0$, then the vector obtained by deleting u_k from $\mathbf{u}$ is an eigenvector of A_k, also with eigenvalue λ_{j+1}. Since this new vector also has exactly j sign changes, it follows, again from Theorem 5.3, that

$$\lambda_{j+1} = \mu_{j+1}^{(k)} < \mu_j^{(k)}.$$

On the other hand, if $w_k = 0$ then $\mathbf{v}' = \mathbf{v}''$ is an eigenvector of A with $j-1$ sign changes. As such

$$\mu_j^{(k)} = \lambda_j > \lambda_{j+1}.$$

Thus we may now assume that both u_k and w_k are nonzero.

The vectors $\mathbf{u}$ and $\mathbf{v}$ are defined up to multiplication by a nonzero constant. As such we assume that u_k and w_k are both positive and set

$$c^* = \inf\{c : c > 0,\ S^-(c\mathbf{v}' + \mathbf{u}) \le j - 1\}.$$

We claim that c^* is a well-defined positive number. This is a consequence of the following. For all c sufficiently large, e.g., such that $c|v_i| > |u_i|$ if $v_i \neq 0$, we have

$$S^-(c\mathbf{v}' + \mathbf{u}) \le S^+(c\mathbf{v}' + \mathbf{u}) \le S^+(\mathbf{v}'') = j - 1.$$

(Here we have used the fact that $cv_k + u_k = u_k$ and w_k are of the same

(i) *There exists an $n \times n$ strictly totally positive matrix A with eigenvalues $\lambda_1 > \cdots > \lambda_n > 0$ and associated eigenvectors $\mathbf{u}^1, \ldots, \mathbf{u}^n$, respectively.*

(ii) *For every $1 \le q \le p \le n$ and nontrivial $(c_q, \ldots, c_p)$*

$$q - 1 \le S^-\left(\sum_{k=q}^{p} c_k \mathbf{u}^k\right) \le S^+\left(\sum_{k=q}^{p} c_k \mathbf{u}^k\right) \le p - 1. \tag{5.5}$$

(iii) *There exist $\varepsilon_p \in \{-1, 1\}$, $p = 1, \ldots, n$, such that*

$$\varepsilon_p U\begin{pmatrix} i_1, \ldots, i_p \\ 1, \ldots, p \end{pmatrix} > 0 \tag{5.6}$$

for all $1 \le i_1 < \cdots < i_p \le n$ and $p = 1, \ldots, n$, and $\delta_q \in \{-1, 1\}$, $q = 1, \ldots, n$, such that

$$\delta_q U\begin{pmatrix} i_q, \ldots, i_n \\ q, \ldots, n \end{pmatrix} (-1)^{\sum_{k=q}^{n} i_k} > 0 \tag{5.7}$$

for all $1 \le i_q < \cdots < i_n \le n$ and $q = 1, \ldots, n$. If we assume that $\varepsilon_n = (-1)^{n(n+1)/2}$ ($\delta_1 = 1$), then $\varepsilon_p \delta_{p+1} = (-1)^{p(p+1)/2}$, $p = 1, \ldots, n$.

Proof (i) ⇒ (ii) is contained in Theorem 5.3.

(ii) ⇒ (iii). We first prove (5.6). The result holds for $p = 1$ by inspection since $S^+(\mathbf{u}^1) = S^-(\mathbf{u}^1) = 0$. We assume, by induction, that the result holds for $p - 1$ and prove it for p, where $p \ge 2$. If

$$U\begin{pmatrix} i_1, \ldots, i_p \\ 1, \ldots, p \end{pmatrix} = 0$$

for some $1 \le i_1 < \cdots < i_p \le n$ then there exists a nontrivial $(c_1, \ldots, c_p)$ such that

$$\left(\sum_{k=1}^{p} c_k \mathbf{u}^k\right)_{i_j} = 0, \qquad j = 1, \ldots, p.$$

But then

$$S^+\left(\sum_{k=1}^{p} c_k \mathbf{u}^k\right) \ge p$$

which is a contradiction. Thus

$$U\begin{pmatrix} i_1, \ldots, i_p \\ 1, \ldots, p \end{pmatrix} \ne 0.$$

It remains to prove that all such minors are of one sign.

Choose $1 \le i_1 < \cdots < i_{p-1} \le n$. Let $(c_1, \ldots, c_p)$ be a nontrivial vector of coefficients satisfying

$$\left(\sum_{k=1}^{p} c_k \mathbf{u}^k\right)_{i_j} = 0, \qquad j = 1, \ldots, p-1.$$

This implies that

$$S^+\left(\sum_{k=1}^{p} c_k \mathbf{u}^k\right) \ge p-1,$$

and from (5.5) we must have

$$S^+\left(\sum_{k=1}^{p} c_k \mathbf{u}^k\right) = p-1.$$

Thus $c_p \neq 0$, and for $i_{j-1} < r < i_j$, where $i_0 = 0$, $i_p = n+1$,

$$\varepsilon(-1)^j \left(\sum_{k=1}^{p} c_k \mathbf{u}^k\right)_r > 0, \qquad j = 1, \ldots, p,$$

for some $\varepsilon \in \{-1, 1\}$. Solve for c_p using the p equations

$$\left(\sum_{k=1}^{p} c_k \mathbf{u}^k\right)_{i_j} = 0, \qquad j = 1, \ldots, p-1,$$

and

$$\left(\sum_{k=1}^{p} c_k \mathbf{u}^k\right)_r = \alpha_r.$$

For any $r \in \{1, \ldots, n\} \backslash \{i_1, \ldots, i_{p-1}\}$, assuming $i_{j-1} < r < i_j$, we have

$$c_p = \frac{\alpha_r U \begin{pmatrix} i_1, \ldots, i_{p-1} \\ 1, \ldots, p-1 \end{pmatrix}}{(-1)^{j+p} U \begin{pmatrix} i_1, \ldots, i_{j-1}, r, i_j, \ldots, i_{p-1} \\ 1, \ldots, p \end{pmatrix}}.$$

As $\operatorname{sgn} \alpha_r = (-1)^j \varepsilon$ for some $\varepsilon \in \{-1, 1\}$ it follows, using the induction hypothesis, that for some $\varepsilon_p \in \{-1, 1\}$

$$\varepsilon_p U \begin{pmatrix} i_1, \ldots, i_{j-1}, r, i_j, \ldots, i_{p-1} \\ 1, \ldots, p \end{pmatrix} > 0$$

for all r between i_{j-1} and i_j and all $j = 1, \ldots, p$, i.e., for all

$$r \in \{1, \ldots, n\} \backslash \{i_1, \ldots, i_{p-1}\}$$

with the rows arranged in increasing order. From a connectedness argument this implies that

$$\varepsilon_p U \begin{pmatrix} i_1, \dots, i_p \\ 1, \dots, p \end{pmatrix} > 0$$

for all $1 \le i_1 < \cdots < i_p \le n$.

We now prove (5.7). For $\mathbf{u} = (u_1, u_2, \dots, u_n)$ we set

$$\widetilde{\mathbf{u}} = (u_1, -u_2, \dots, (-1)^{n+1} u_n)$$

and recall that

$$S^-(\mathbf{u}) + S^+(\widetilde{\mathbf{u}}) = n - 1$$

for every $\mathbf{u} \in \mathbb{R}^n$, $\mathbf{u} \neq \mathbf{0}$, see Lemma 3.1. Thus the inequality

$$S^- \left(\sum_{k=q}^{n} c_k \mathbf{u}^k \right) \ge q - 1,$$

is equivalent to

$$S^+ \left(\sum_{k=q}^{n} c_k \widetilde{\mathbf{u}}^k \right) \le n - q,$$

for nontrivial $(c_q, \dots, c_n)$. This is exactly the same inequality as assumed in the previous paragraph. As such there exists a $\widetilde{\delta}_q \in \{-1, 1\}$ such that

$$\widetilde{\delta}_q \widetilde{U} \begin{pmatrix} i_q, \dots, i_n \\ q, \dots, n \end{pmatrix} > 0$$

for all $1 \le i_q < \cdots < i_n \le n$, where $\widetilde{U}$ denotes the $n \times n$ matrix with columns $\widetilde{\mathbf{u}}^k$. This immediately translates into

$$\widetilde{\delta}_q U \begin{pmatrix} i_q, \dots, i_n \\ q, \dots, n \end{pmatrix} (-1)^{\sum_{k=q}^{n} i_k + 1} > 0,$$

implying

$$\delta_q U \begin{pmatrix} i_q, \dots, i_n \\ q, \dots, n \end{pmatrix} (-1)^{\sum_{k=q}^{n} i_k} > 0$$

for some $\delta_q \in \{-1, 1\}$ and all $1 \le i_q < \cdots < i_n \le n$.

Multiplying any one of the $\mathbf{u}^k$ by a nonzero constant we may and will assume that

$$U \begin{pmatrix} 1, \dots, n \\ 1, \dots, n \end{pmatrix} = (-1)^{n(n+1)/2},$$

i.e., $\varepsilon_n = (-1)^{n(n+1)/2}$ and $\delta_1 = 1$. Now for any $p \in \{1, \ldots, n-1\}$, by the Laplace expansion by minors,

$$\begin{aligned}(-1)^{n(n+1)/2} &= \det U \\ &= \sum_{1 \le i_1 < \cdots < i_p \le n} (-1)^{\sum_{k=p+1}^n i'_k + k} U\begin{pmatrix} i_1, \ldots, i_p \\ 1, \ldots, p \end{pmatrix} U \begin{pmatrix} i'_{p+1}, \ldots, i'_n \\ p+1, \ldots, n \end{pmatrix}\end{aligned}$$

where the $i_1 < \cdots < i_p$ and $i'_{p+1} < \cdots < i'_n$ are complementary indices in $\{1, \ldots, n\}$. As

$$\varepsilon_p U \begin{pmatrix} i_1, \ldots, i_p \\ 1, \ldots, p \end{pmatrix} > 0$$

and

$$\delta_{p+1} U \begin{pmatrix} i'_{p+1}, \ldots, i'_n \\ p+1, \ldots, n \end{pmatrix} (-1)^{\sum_{k=p+1}^n i'_k} > 0$$

it follows that $\varepsilon_p \delta_{p+1} = (-1)^{p(p+1)/2}$.

(iii) $\Rightarrow$ (i). Set $V = U^{-1}$ where, as above, we assume, without loss of generality, that $\det U = (-1)^{n(n+1)/2}$. Then

$$V \begin{pmatrix} 1, \ldots, p \\ j_1, \ldots, j_p \end{pmatrix} = (-1)^{\sum_{k=1}^p k + j_k} U \begin{pmatrix} j'_1, \ldots, j'_{n-p} \\ p+1, \ldots, n \end{pmatrix} (-1)^{n(n+1)/2}$$

where the $j_1 < \cdots < j_p$ and $j'_1 < \cdots < j'_{n-p}$ are complementary indices in $\{1, \ldots, n\}$. As

$$\delta_{p+1} U \begin{pmatrix} j'_1, \ldots, j'_{n-p} \\ p+1, \ldots, n \end{pmatrix} (-1)^{\sum_{k=1}^{n-p} j'_k} > 0$$

we have

$$\delta_{p+1} (-1)^{p(p+1)/2} V \begin{pmatrix} 1, \ldots, p \\ j_1, \ldots, j_p \end{pmatrix} > 0$$

for all $1 \le j_1 < \cdots < j_p \le n$ and all $p = 1, \ldots, n-1$, i.e.,

$$\varepsilon_p V \begin{pmatrix} 1, \ldots, p \\ j_1, \ldots, j_p \end{pmatrix} > 0$$

for all $1 \le j_1 < \cdots < j_p \le n$ and all $p = 1, \ldots, n$. This implies that

$$U \begin{pmatrix} i_1, \ldots, i_p \\ 1, \ldots, p \end{pmatrix} V \begin{pmatrix} 1, \ldots, p \\ j_1, \ldots, j_p \end{pmatrix} > 0 \tag{5.8}$$

for every choice of $1 \le i_1 < \cdots < i_p \le n$ and $1 \le j_1 < \cdots < j_p \le n$, and all p.

For any $\lambda_1 > \cdots > \lambda_n > 0$ let Λ be the $n \times n$ diagonal matrix with diagonal entries $\{\lambda_1, \ldots, \lambda_n\}$. Consider

$$A = U\Lambda V.$$

The columns of U are the (right) eigenvectors of A with associated eigenvalues $\lambda_1, \ldots, \lambda_n$. We will choose the λ_j dependent upon U and V.

To prove that A is strictly totally positive it suffices to show that

$$A\begin{pmatrix} i_1, \ldots, i_p \\ j_1, \ldots, j_p \end{pmatrix} > 0$$

for all choices of $1 \le i_1 < \cdots < i_p \le n$ and $1 \le j_1 < \cdots < j_p \le n$. For A as above we have from the Cauchy–Binet formula

$$A\begin{pmatrix} i_1, \ldots, i_p \\ j_1, \ldots, j_p \end{pmatrix} = \sum_{1 \le k_1 < \cdots < k_p \le n} U\begin{pmatrix} i_1, \ldots, i_p \\ k_1, \ldots, k_p \end{pmatrix} \lambda_{k_1} \cdots \lambda_{k_p} V\begin{pmatrix} k_1, \ldots, k_p \\ j_1, \ldots, j_p \end{pmatrix}.$$

From (5.8) we see that we can choose the $\lambda_1 > \cdots > \lambda_n > 0$ so that A is strictly totally positive. (From the above construction we see that for any choice of $\lambda_1 > \cdots > \lambda_n > 0$ and $A = U\Lambda V$, we have that $A^m = U\Lambda^m V$ is a strictly totally positive matrix for m sufficiently large.) □

5.5 Eigenvalues as functions of matrix elements

How do the eigenvalues of an oscillation matrix vary as functions of the elements of the matrix? This is the topic of this section.

We start with a general result.

Proposition 5.9 *Let $A = (a_{ij})$ be an $n \times n$ matrix with distinct simple eigenvalues $\lambda_1, \ldots, \lambda_n$ and associated eigenvectors $\mathbf{u}^1, \ldots, \mathbf{u}^n$. Let $U = (u_{ij})$ be the $n \times n$ eigenmatrix whose kth column is $\mathbf{u}^k$. Set $V = (v_{ij})$ where $V = U^{-1}$, and let $\mathbf{v}^k$ denote the kth row of V. Then*

$$\frac{\partial \lambda_k}{\partial a_{ij}} = v_{ki} u_{jk}$$

for all $i, j, k \in \{1, \ldots, n\}$.

Proof The $\{\lambda_k\}$ are locally differentiable functions of the a_{ij} since they are the distinct roots of a polynomial of degree n whose coefficients depend algebraically upon the a_{ij}.

From the above $A\mathbf{u}^k = \lambda_k \mathbf{u}^k$, and thus

$$\sum_{s=1}^{n} a_{rs}u_{sk} = \lambda_k u_{rk}, \qquad r = 1, \ldots, n.$$

Similarly $\mathbf{v}^k A = \lambda_k \mathbf{v}^k$, i.e.,

$$\sum_{r=1}^{n} v_{kr}a_{rs} = \lambda_k v_{ks}, \qquad s = 1, \ldots, n.$$

Now

$$A = U\Lambda V$$

where Λ is the $n \times n$ diagonal matrix with diagonal entries $\{\lambda_1, \ldots, \lambda_n\}$. Thus

$$VAU = \Lambda,$$

i.e.,

$$\lambda_k = \sum_{r,s=1}^{n} v_{kr}a_{rs}u_{sk}, \qquad k = 1, \ldots, n.$$

Differentiating the above with respect to a_{ij}, we have

$$\begin{aligned} \frac{\partial \lambda_k}{\partial a_{ij}} &= v_{ki}u_{jk} + \sum_{r,s=1}^{n} \frac{\partial v_{kr}}{\partial a_{ij}} a_{rs}u_{sk} + v_{kr}a_{rs}\frac{\partial u_{sk}}{\partial a_{ij}} \\ &= v_{ki}u_{jk} + \sum_{r=1}^{n} \frac{\partial v_{kr}}{\partial a_{ij}} \left(\sum_{s=1}^{n} a_{rs}u_{sk} \right) + \sum_{s=1}^{n} \left(\sum_{r=1}^{n} v_{kr}a_{rs} \right) \frac{\partial u_{sk}}{\partial a_{ij}}. \end{aligned}$$

Substituting from above gives

$$\begin{aligned} \frac{\partial \lambda_k}{\partial a_{ij}} &= v_{ki}u_{jk} + \sum_{r=1}^{n} \frac{\partial v_{kr}}{\partial a_{ij}} \lambda_k u_{rk} + \sum_{s=1}^{n} \lambda_k v_{ks} \frac{\partial u_{sk}}{\partial a_{ij}} \\ &= v_{ki}u_{jk} + \lambda_k \left(\sum_{r=1}^{n} \frac{\partial v_{kr}}{\partial a_{ij}} u_{rk} + \sum_{s=1}^{n} v_{ks}\frac{\partial u_{sk}}{\partial a_{ij}} \right) \\ &= v_{ki}u_{jk} + \lambda_k \frac{\partial}{\partial a_{ij}} \left(\sum_{r=1}^{n} v_{kr}u_{rk} \right) \\ &= v_{ki}u_{jk} \end{aligned}$$

since $\sum_{r=1}^{n} v_{kr}u_{rk} \equiv 1$ as $V = U^{-1}$. □

An application of Proposition 5.9 leads to the following.

Theorem 5.10 *Let $A = (a_{ij})$ be an $n \times n$ oscillation matrix with eigenvalues $\lambda_1 > \cdots > \lambda_n > 0$. Then*

$$\frac{\partial \lambda_1}{\partial a_{ij}} > 0, \ (-1)^{i+j}\frac{\partial \lambda_n}{\partial a_{ij}} > 0, \qquad i,j = 1,\ldots,n,$$

and

$$\frac{\partial \lambda_k}{\partial a_{11}} > 0, \ \frac{\partial \lambda_k}{\partial a_{nn}} > 0, \ (-1)^{k+1}\frac{\partial \lambda_k}{\partial a_{1n}} > 0, \ (-1)^{k+1}\frac{\partial \lambda_k}{\partial a_{n1}} > 0, \ k = 1,\ldots,n.$$

Proof Let U be as in Proposition 5.9. Without loss of generality we assume that $u_{1k} > 0$, $k = 1,\ldots,n$. This implies, from Theorem 5.3, that $u_{k1} > 0$, $k = 1,\ldots,n$, and $(-1)^{k+1}u_{kn} > 0$, $(-1)^{k+1}u_{nk} > 0$, $k = 1,\ldots,n$. Set $V = U^{-1}$. Then the exact same inequalities hold for V, i.e., $v_{1k}, v_{k1} > 0$, $k = 1,\ldots,n$, and $(-1)^{k+1}v_{kn}, (-1)^{k+1}v_{nk} > 0$, $k = 1,\ldots,n$. Theorem 5.10 is now an immediate consequence of these inequalities and the previous Proposition 5.9. □

The inequality

$$\frac{\partial \lambda_1}{\partial a_{ij}} > 0$$

is also a consequence of Perron's Theorem. Another consequence of Perron's Theorem is the following.

Let

$$A_c = \begin{pmatrix} a_{11} & \cdots & a_{1n} \\ \vdots & & \vdots \\ ca_{k1} & \cdots & ca_{kn} \\ \vdots & & \vdots \\ a_{n1} & \cdots & a_{nn} \end{pmatrix}$$

i.e., A_c is obtained from A by multiplying the elements of the kth row by c.

Proposition 5.11 *Let A be an $n \times n$ oscillation matrix. For $c > 0$, let $\lambda_1(c) > \cdots > \lambda_n(c) > 0$ denote the eigenvalues of A_c . Then*

$$\frac{\partial \lambda_1(c)}{\partial c} > 0, \quad \frac{\partial \lambda_n(c)}{\partial c} > 0$$

and

$$\frac{\partial(\lambda_1(c)\cdots\lambda_r(c))}{\partial c} > 0, \qquad r = 1,\ldots,n.$$

Proof All three inequalities are consequences of Perron's Theorem. That is, we use the characterization

$$\lambda_1 = \sup\{\lambda : A\mathbf{x} \geq \lambda\mathbf{x}, \text{ for some } \mathbf{x} \geq \mathbf{0}\}$$

valid for any strictly positive matrix. The first inequality is a direct application thereof (and is also in Theorem 5.10). The second inequality follows from an application of Perron's Theorem to the oscillation matrix $DA_c^{-1}D$. For a proof of the third inequality, apply Perron's Theorem to the rth compound matrix $A_{[r]}$. □

It is not necessarily true that

$$\frac{\partial \lambda_r(c)}{\partial c} > 0$$

for all r. The 3×3 totally positive matrix

$$A_c = \begin{pmatrix} 1 & 1 & 0 \\ 2c & 2c & c \\ 2 & 2 & 1 \end{pmatrix}$$

is such that $\lambda_2(c)$ is a strictly decreasing function of c, $c > 0$.

5.6 Remarks

The study of the spectral properties of integral equations with totally positive continuous kernels substantially predates the study of the spectral properties of totally positive matrices. In 1918 O. D. Kellogg proved the main spectral properties in the case of a *symmetric* continuous totally positive kernel (see Kellogg [1918]). Kellogg was an American mathematician who obtained his doctorate from Göttingen in 1903 under the supervision of Hilbert. He is best known for his work on potential theory, and his book thereon Kellogg [1929] has been reprinted many times since. Both Krein and Gantmacher were very much influenced by Kellogg's work on this subject. An announcement of the parallel result for continuous *non-symmetric* totally positive kernels is in Gantmacher [1936].

The main results concerning spectral properties of totally positive matrices are in Gantmacher, Krein [1937]. An announcement appears in Gantmacher, Krein [1935]. Oscillation matrices were introduced in Gantmacher, Krein [1937], and Theorem 5.2 can be found therein on p. 454 (see also Gantmacher, Krein [1950], Chap. II, §7, Karlin [1968], Chap. 2, §9, and Ando [1987]). The proof of Theorem 5.3, based on the Perron and Kronecker theorems, is from Gantmacher, Krein [1937], Theorems 10 and 14. The same proof also appears in Gantmacher [1953], Gantmacher, Krein [1950], and Ando [1987]. The strict interlacing of the eigenvalues of a strictly totally positive matrix with the eigenvalues of the principal submatrices obtained by deleting the first (or last) row and column was first proved in Gantmacher, Krein [1937]. The proof as given in Proposition 5.4 is from Koteljanskii [1955]. The second proof of Theorem 5.3, based on variation diminishing, is from Elias, Pinkus [2002], which contains generalizations of Theorem 5.3 to the nonlinear setting (see also Pinkus [1985a], [1985b] and Buslaev [1990]). For more on the eigenvalues and eigenvectors of oscillation matrices see Eveson [1996], Karlin [1965], Karlin [1972] and Karlin, Pinkus [1974]. Fallat, Gekhtman, Johnson [2000] consider the possible spectral structure of a subclass of totally positive matrices. Theorem 5.6 is due to Pinkus [1998]. The example of a totally positive matrix where interlacing of the eigenvalues of the minor fails is from Karlin, Pinkus [1974]. Friedland [1985] had previously proved that $\mu_1^{(k)} \ge \lambda_2$ and $\mu_{n-1}^{(k)} \ge \lambda_n$ for $k = 1, \ldots, n-1$. Parts of Theorem 5.8, in a slightly different form, appear in Gantmacher, Krein [1937], Theorem 16. Theorem 5.10 is also from Gantmacher, Krein [1937], Theorems 18 and 19.

A survey of the spectral properties of totally positive kernels and matrices, with extensive references, can be found in Pinkus [1996].

6

Factorizations of totally positive matrices

In Chapter 2 (Theorem 2.12) we proved that every $n \times n$ totally positive matrix can be factored in the form

$$A = LU$$

where L and U are totally positive matrices, L being a lower triangular matrix and U an upper triangular matrix.

In this chapter we study factorizations of totally positive matrices in considerably greater detail. We prove that if A is an $n \times n$ totally positive matrix, then A can always be factored in the form

$$A = L^1 \cdots L^{n-1} U^1 \cdots U^{n-1}$$

where each L^k is a 1-banded, lower triangular, totally positive matrix and each U^k is a 1-banded, upper triangular, totally positive matrix. In fact there are generally an uncountable number of such factorizations, and at least 2^{2n-2} different factorizations with a maximum number of zero entries.

In Section 6.2, after some preliminaries, we start with one such factorization for A strictly totally positive by essentially writing down the factorization, and then proving its validity. We then show, using elementary operations, how it is possible to construct many other such factorizations. In Section 6.3 we generalize these results to totally positive matrices.

6.1 Preliminaries

We start with two definitions to set our notation.

Definition 6.1 For each $i, j \in \{1, \ldots, n\}$, $i \neq j$, we define

$$E_{i,j}(\alpha)$$

as the $n \times n$ unit diagonal matrix with α in the (i, j) position and zeroes in the other off-diagonal entries.

We recall that right multiplication of A by $E_{i,j}(\alpha)$ is the operation whereby α times the ith column of A is added to the jth column of A with the other columns left unchanged. Left multiplication of A by $E_{i,j}(\alpha)$ is the operation whereby α times the jth row of A is added to the ith row of A with the other rows left unchanged. In addition, as is readily verified, $(E_{ij}(\alpha))^{-1} = E_{ij}(-\alpha)$.

Definition 6.2 We say that a lower triangular matrix $L = (\ell_{ij})$ is r-banded if $\ell_{ij} = 0$ for $i - j > r$.

Thus L has the form

$$L = \begin{pmatrix} 1 & & & & & & \\ \times & \cdot & & & & & \\ \times & \cdot & \cdot & & & & \\ \times & \cdot & \cdot & \cdot & & & \\ \times & \cdot & \cdot & \cdot & \cdot & & \\ & \cdot & \cdot & \cdot & \cdot & \cdot & \\ & & \cdot & \cdot & \cdot & \cdot & \cdot \\ & & & \underbrace{\times \quad \times \quad \times \quad \times}_{r} & & & 1 \end{pmatrix}.$$

If L is an $n \times n$ matrix and, for whatever reason, we say that L is r-banded and lower triangular, where $r \geq n - 1$, then this statement is meaningless and should be so understood.

We list three elementary properties of r-banded matrices for easy reference.

Proposition 6.3 *If L is an r-banded lower triangular matrix and M is an s-banded lower triangular matrix, then LM is an $(r + s)$-banded lower triangular matrix.*

Proposition 6.4 *A 1-banded lower triangular matrix is totally positive if and only if all its elements are nonnegative.*

We repeatedly use the following (contrast it with Proposition 6.3).

Proposition 6.5 *For $k = 1, \ldots, n - 1$, let*

$$C^k = E_{2,1}(\alpha_1) \cdots E_{k+1,k}(\alpha_k).$$

Then C^k is a 1-banded unit diagonal lower triangular matrix with

$$\begin{aligned}\left(C^k\right)_{i+1,i} &= \alpha_i, \qquad i = 1,\ldots,k,\\ \left(C^k\right)_{i+1,i} &= 0, \qquad i = k+1,\ldots,n-1.\end{aligned}$$

If

$$R^k = E_{n-k+1,n-k}(\alpha_{n-k})\cdots E_{n,n-1}(\alpha_{n-1}),$$

then R^k is a 1-banded unit diagonal lower triangular matrix with

$$\begin{aligned}\left(R^k\right)_{i+1,i} &= 0, \qquad i = 1,\ldots,n-k-1,\\ \left(R^k\right)_{i+1,i} &= \alpha_i, \qquad i = n-k,\ldots,n-1.\end{aligned}$$

Note that when reversing the multiplication order (in C^k) we get

$$E_{k+1,k}(\alpha_1)\cdots E_{2,1}(\alpha_k),$$

which is a k-banded matrix, whose (i,j)th entry, $1 \le j < i \le k$, is exactly

$$\alpha_j \cdots \alpha_{i-1},$$

with other off-diagonal entries being zero.

6.2 Factorizations of strictly totally positive matrices

We recall (Theorem 2.10) that an $n \times n$ strictly totally positive matrix A has a unique factorization of the form

$$A = LDU$$

where L is a unit diagonal lower strictly totally positive matrix, U is a unit diagonal upper strictly totally positive matrix, and D is a diagonal matrix whose diagonal entries are strictly positive. We also recall that the unit diagonal matrix L is lower strictly totally positive if it is lower triangular and

$$L\begin{pmatrix} i_1,\ldots,i_k \\ j_1,\ldots,j_k \end{pmatrix} > 0$$

whenever $i_m \ge j_m$, $m = 1,\ldots,k$.

We first prove a fundamental result from which all other factorization results will follow.

Theorem 6.6 *Let L be an $n \times n$ unit diagonal lower triangular matrix.*

Then L is lower strictly totally positive if and only if L can be factored in the form

$$L = R^1 \cdots R^{n-1} \tag{6.1}$$

where

$$R^k = E_{n-k+1,n-k}(\alpha_{k,n-k}) \cdots E_{n,n-1}(\alpha_{k,n-1}), \qquad k = 1, \ldots, n-1, \tag{6.2}$$

with $\alpha_{k,j} > 0$, $j = n-k, \ldots, n-1$, $k = 1, \ldots, n-1$.

Proof Assume L has the form (6.1) where each R^k satisfies (6.2), with the appropriate $\alpha_{k,j}$ strictly positive. As L is the product of unit diagonal lower totally positive matrices, it is necessarily a unit diagonal lower totally positive matrix. But we want to prove that it is a lower strictly totally positive matrix. We recall (Theorem 2.8) that to prove that L is lower strictly totally positive it suffices to prove that

$$L\begin{pmatrix} i+1, \ldots, i+r \\ 1, \ldots, r \end{pmatrix} > 0, \qquad i = 0, 1, \ldots, n-r,\ r = 1, \ldots, n.$$

(Since L is unit diagonal, the equations in the case $i = 0$ are of no interest.) In fact (see Proposition 2.9), we need only prove the strict positivity of the above inequalities for $i = n-r$. However, in the proof of the converse direction we make use of all these inequalities, and so we also prove them here.

Recall that

$$R^k = \begin{pmatrix} 1 & & & & & & \\ 0 & \cdot & & & & & \\ & \cdot & \cdot & & & & \\ & & \cdot & \cdot & & & \\ & & & 0 & \cdot & & \\ & & & & \alpha_{k,n-k} & \cdot & \\ & & & & & \cdot & \cdot \\ & & & & & & \cdot \quad \cdot \\ & & & & & & \alpha_{k,n-1} \quad 1 \end{pmatrix}.$$

Because of the format (zero elements) of the R^k it easily follows that

$$(L)_{i+1,1} = (R^{n-i})_{i+1,i} \cdots (R^{n-1})_{2,1}, \qquad i = 1, \ldots, n-1.$$

As

$$(R^{n-i})_{i+1,i} > 0, \qquad i = 1, \ldots, n-1,$$

we have

$$(L)_{i+1,1} > 0, \qquad i = 1, \ldots, n-1.$$

Now consider

$$L\begin{pmatrix} i+1, i+2 \\ 1, 2 \end{pmatrix}.$$

Again, from the form of the $R^1, \ldots, R^{n-1}$, it follows that

$$L\begin{pmatrix} i+1, i+2 \\ 1, 2 \end{pmatrix} = R^{n-i}\begin{pmatrix} i+1, i+2 \\ i, i+1 \end{pmatrix} \cdots R^{n-1}\begin{pmatrix} 2, 3 \\ 1, 2 \end{pmatrix}, \quad i = 1, \ldots, n-2.$$

By assumption,

$$R^{n-i}\begin{pmatrix} i+1, i+2 \\ i, i+1 \end{pmatrix} = (R^{n-i})_{i+1,i}(R^{n-i})_{i+2,i+1} > 0, \quad i = 1, \ldots, n-2.$$

Thus

$$L\begin{pmatrix} i+1, i+2 \\ 1, 2 \end{pmatrix} > 0, \qquad i = 1, \ldots, n-2.$$

We progress in this fashion to obtain

$$L\begin{pmatrix} i+1, \ldots, i+r \\ 1, \ldots, r \end{pmatrix} > 0, \qquad i = 1, \ldots, n-r, \; r = 1, \ldots, n-1.$$

Thus L is lower strictly totally positive.

Let us now assume that L is lower strictly totally positive. For $k = 1, \ldots, n-1$ we define the unit diagonal, 1-banded matrix R^k with

$$\begin{aligned} (R^k)_{i+1,i} &= 0, \qquad i = 1, \ldots, n-k-1, \\ (R^k)_{i+1,i} &> 0, \qquad i = n-k, \ldots, n-1 \end{aligned}$$

in the following manner. We will determine the $(R^k)_{i+1,i}$, $i = n-k, \ldots, n-1$, $k = 1, \ldots, n-1$, from the previous sets of equations. That is, the equations

$$(L)_{i+1,1} = (R^{n-i})_{i+1,i} \cdots (R^{n-1})_{2,1}, \qquad i = 1, \ldots, n-1,$$

and the fact that $(L)_{i+1,i} > 0$ define for us the strictly positive values

$$(R^{n-i})_{i+1,i}, \qquad i = 1, \ldots, n-1.$$

The equations

$$L\begin{pmatrix} i+1, i+2 \\ 1, 2 \end{pmatrix} = R^{n-i}\begin{pmatrix} i+1, i+2 \\ i, i+1 \end{pmatrix} \cdots R^{n-1}\begin{pmatrix} 2, 3 \\ 1, 2 \end{pmatrix}, \quad i = 1, \ldots, n-2,$$

and the strict positivity of the left-hand side of the equations define for us the strictly positive values

$$R^{n-i}\begin{pmatrix} i+1, i+2 \\ i, i+1 \end{pmatrix} = (R^{n-i})_{i+1,i}(R^{n-i})_{i+2,i+1}, \quad i = 1, \ldots, n-2,$$

(since we set $(R^{n-i})_{i+2,i} = 0$), which in turn define for us the strictly positive values

$$(R^{n-i})_{i+2,i+1}, \qquad i = 1, \ldots, n-2.$$

We continue in this fashion to define all the desired $(R^k)_{i+1,i}$.

We claim that

$$L = R^1 \cdots R^{n-1}.$$

That is, we have chosen the values of the R^k so that the formulæ for the

$$L\begin{pmatrix} i+1, \ldots, i+r \\ 1, \ldots, r \end{pmatrix}, \qquad i = 1, \ldots, n-r,\ r = 1, \ldots, n-1,$$

are valid. But do these formulæ necessarily guarantee that

$$L = R^1 \cdots R^{n-1}?$$

The answer is, of course, yes. The above set of data, and their strict positivity, determine explicitly all the entries of L. That is, the first column of L is given ($r = 1$). Given the first column (and the fact that its entries are not zero) and

$$L\begin{pmatrix} i+1, i+2 \\ 1, 2 \end{pmatrix}, \qquad i = 1, \ldots, n-2,$$

determines the entries of the second column of L, etc... Thus we do have

$$L = R^1 \cdots R^{n-1}.$$

This proves the theorem. □

Remark From the above follow very explicit formulæ for the $(R^k)_{i+1,i}$, $i = n-k, \ldots, n-1$, $k = 1, \ldots, n-1$, in terms of minors of L. But we shall not list them here.

Let us now consider how we can construct a different factorization for L.

Corollary 6.7 *Let L be an $n \times n$ unit diagonal lower triangular matrix. Then L is lower strictly totally positive if and only if L can be factored in the form*

$$L = C^{n-1} \cdots C^1 \tag{6.3}$$

where

$$C^k = E_{2,1}(\beta_{k,1}) \cdots E_{k+1,k}(\beta_{k,k}) \tag{6.4}$$

with $\beta_{k,j} > 0$, $j = 1, \ldots, k$, $k = 1, \ldots, n-1$.

Proof There are two simple ways of justifying this corollary. The first is to take the proof of Theorem 6.6 and make the simple obvious modifications therein.

The second explanation is more satisfying. Recall from Propositions 1.2 and 1.3 that a matrix A is totally positive (strictly totally positive) if and only if A^T is totally positive (strictly totally positive) if and only if QAQ is totally positive (strictly totally positive), where

$$Q = \begin{pmatrix} 0 & \cdots & 0 & 1 \\ 0 & \cdots & 1 & 0 \\ \vdots & \ddots & \vdots & \vdots \\ 1 & \cdots & 0 & 0 \end{pmatrix}.$$

Note that QAQ is the matrix A with the order of its rows and columns reversed. Given a lower triangular matrix $L = (\ell_{ij})_{i,j=1}^n$, then

$$(QLQ)^T = QL^TQ = (\ell_{n-j+1,n-i+1})_{i,j=1}^n$$

is also lower triangular. In addition, L is lower strictly totally positive if and only if $(QLQ)^T$ is lower strictly totally positive.

Applying Theorem 6.6 to $(QLQ)^T$ gives us

$$(QLQ)^T = R^1 \cdots R^{n-1}$$

for $R^1, \ldots, R^{n-1}$ as in the statement of Theorem 6.6. Thus

$$L = (QR^{n-1}Q)^T \cdots (QR^1Q)^T.$$

Set

$$C^k = (QR^kQ)^T, \qquad k = 1, \ldots, n-1.$$

The C^k are as in the statement of Corollary 6.7. □

We can and should look at the factorizations of Theorem 6.6 and Corollary 6.7 from a somewhat different (and historically more valid) perspective. To explain, let

$$S^k = R^k \cdots R^{n-1}, \qquad k = 1, \ldots, n-1.$$

Thus

$$L = R^1 \cdots R^{k-1} S^k.$$

As may be seen, S^k is a full $(n-k)$-banded matrix (more on the structure of the S^k anon) and

$$S^k = R^k S^{k+1}.$$

To explain what is happening recall that $(E_{i,j}(\alpha))^{-1} = E_{i,j}(-\alpha)$ and

$$R^k = E_{n-k+1,n-k}(\alpha_{k,n-k}) \cdots E_{n,n-1}(\alpha_{k,n-1}).$$

Thus

$$(R^k)^{-1} = E_{n,n-1}(-\alpha_{k,n-1}) \cdots E_{n-k+1,n-k}(-\alpha_{k,n-k}),$$

and

$$S^{k+1} = E_{n,n-1}(-\alpha_{k,n-1}) \cdots E_{n-k+1,n-k}(-\alpha_{k,n-k}) S^k.$$

As previously noted, left multiplication of a matrix A by $E_{r+1,r}(-\alpha)$ is the operation whereby α times the rth row of A is subtracted from the $r+1$st row of A, with the other rows left unchanged. Thus S^{k+1} is obtained from S^k by successively eliminating (making zero) the k strictly positive elements along the $(n-k)$-band of S^k (i.e., elements in positions $(n-k+j, j)$, $j = 1, \ldots, k$). It does this by successive row subtraction. It first eliminates the element in the $(n-k+1, 1)$ position by subtracting $\alpha_{k,n-k}$ times the $(n-k)$th row from the $(n-k+1)$st row. It then eliminates the element in the $(n-k+2, 2)$ position by subtracting $\alpha_{k,n-k+1}$ times the $(n-k+1)$st row from the $(n-k+2)$nd row. This does not effect the previous "elimination," etc.

Similarly, from

$$L = C^{n-1} \cdots C^1$$

as in Corollary 6.7, let

$$T^k = C^{n-1} \cdots C^k, \qquad k = 1, \ldots, n-1.$$

Thus

$$L = T^k C^{k-1} \cdots C^1$$

and

$$T^k = T^{k+1} C^k.$$

The matrix T^k is a full $(n-k)$-banded matrix. The difference here is that $(C^k)^{-1}$ uses column subtraction (rather than the row subtraction of

$(R^k)^{-1}$) to take the $(n-k)$-banded matrix T^k to the $(n-k-1)$-banded matrix T^{k+1}.

Starting from a unit diagonal lower strictly totally positive matrix it is readily verified that both the S^k and T^k are unit diagonal lower totally positive matrices and are as strictly totally positive as they can be, considering their geometric structure, i.e., considering the fact that they are $(n-k)$-banded. The matrix S^k satisfies

$$S^k \begin{pmatrix} i_1, \ldots, i_r \\ j_1, \ldots, j_r \end{pmatrix} > 0$$

if and only if

$$j_s \le i_s \le j_s + (n-k), \qquad s = 1, \ldots, r,$$

and the exact same inequalities hold for T^k. Thus the S^k and T^k have exactly the same total positivity structure.

We can state this procedure formally as follows.

Proposition 6.8 *Let P be an $n \times n$ unit diagonal lower triangular $(n-k)$-banded totally positive matrix for which*

$$P \begin{pmatrix} i_1, \ldots, i_r \\ j_1, \ldots, j_r \end{pmatrix} > 0$$

if and only if $j_s \le i_s \le j_s + (n-k)$, $s = 1, \ldots, r$. Then there exist R^k and C^k where

$$R^k = E_{n-k+1,n-k}(\alpha_{n-k}) \cdots E_{n,n-1}(\alpha_{n-1})$$

with $\alpha_j > 0$, $j = n-k, \ldots, n-1$, and

$$C^k = E_{2,1}(\beta_1) \cdots E_{k+1,k}(\beta_k)$$

with $\beta_j > 0$, $j = 1, \ldots, k$, such that

$$P = R^k S = T C^k$$

where S and T are $n \times n$ unit diagonal lower triangular $(n-k-1)$-banded totally positive matrices and

$$S \begin{pmatrix} i_1, \ldots, i_r \\ j_1, \ldots, j_r \end{pmatrix}, \quad T \begin{pmatrix} i_1, \ldots, i_r \\ j_1, \ldots, j_r \end{pmatrix} > 0$$

if and only if $j_s \le i_s \le j_s + (n-k-1)$, $s = 1, \ldots, r$.

We sequentially did the same thing at each step in Theorem 6.6 and 6.7, i.e., we decomposed

$$S^k = R^k S^{k+1}$$

for $k = 1, \ldots, n-1$, in Theorem 6.6 and decomposed

$$T^k = T^{k+1} C^k$$

for $k = 1, \ldots, n-1$, in Corollary 6.7. However we could just as easily have decomposed S^k via

$$S^k = \widehat{S}^{k+1} \widehat{C}^k$$

and T^k via

$$T^k = \widehat{R}^k \widehat{T}^{k+1}$$

where $\widehat{C}^k$ has the structure of C^k, $\widehat{R}^k$ has the structure of R^k, and $\widehat{S}^{k+1}$, $\widehat{T}^{k+1}$, S^{k+1} and T^{k+1} all have the same structure. This is the content of Proposition 6.8. Since we can make the choice of using row subtraction (R^k) or column subtraction (C^k) at each step of the procedure we see that we actually have 2^{n-1} possible factorizations. (Note that there is an ordering to these 2^{n-1} factorizations. Consider $R^1 \cdots R^{n-1} C^{n-1} \cdots C^1$ and for each k delete one and only one of the R^k or C^k. These are the possible forms of the factorizations where it is understood that in every distinct factorization the entries of the R^k or C^k depend on the choice of the factorization.)

What characterizes these 2^{n-1} factorizations is that the R^k and C^k have exactly k positive off-diagonal entries, the last k entries or the first k entries, respectively, and this is minimal. On the other hand, without the minimality of nonzero terms there can be (and will be) a continuum of factorizations of a lower strictly totally positive $n \times n$ matrix into the product of $(n-1)$ 1-banded totally positive factors. For example, consider the matrix

$$A = \begin{pmatrix} 1 & 0 & 0 \\ 1 & 1 & 0 \\ 1 & 2 & 1 \end{pmatrix}.$$

This matrix can be factored into a continuum of a product of two 1-banded unit diagonal lower totally positive matrices, namely for every $x \in [1/2, 1]$

$$A = \begin{pmatrix} 1 & 0 & 0 \\ 1 & 1 & 0 \\ 1 & 2 & 1 \end{pmatrix} = \begin{pmatrix} 1 & 0 & 0 \\ 1-x & 1 & 0 \\ 0 & 1/x & 1 \end{pmatrix} \begin{pmatrix} 1 & 0 & 0 \\ x & 1 & 0 \\ 0 & 2-1/x & 1 \end{pmatrix}.$$

(Every 3×3 lower strictly totally positive matrix can be factored into a continuum of products of two 1-banded lower totally positive matrices.)

What was done for a lower strictly totally positive matrix is also valid for an upper strictly totally positive matrix, with the obvious modifications. Thus, taking into consideration the above results together with the LDU factorization of a strictly totally positive matrix as detailed in Theorem 2.10, we can summarize as follows.

Theorem 6.9 *Let A be an $n \times n$ strictly totally positive matrix. Then A can be factored in many ways in the form*

$$A = L^1 \cdots L^{n-1} D U^1 \cdots U^{n-1}$$

where D is a diagonal matrix whose diagonal entries are strictly positive, the L^k are 1-banded, unit diagonal, lower totally positive matrices, and the U^k are 1-banded, unit diagonal, upper totally positive matrices. There are also many (in general 2^{2n-2}) different factorizations of this form where among the $L^1, \ldots, L^{n-1}$ there is exactly one matrix with k strictly positive entries on its off-diagonal, $k = 1, \ldots, n-1$, and among the $U^1, \ldots, U^{n-1}$ there is exactly one matrix with k strictly positive entries on its off-diagonal, $k = 1, \ldots, n-1$.

We should also recall that every strictly totally positive (totally positive matrix) also has a factorization of the form $A = UDL$, where U is upper strictly totally positive (totally positive) and L is lower strictly totally positive (totally positive). Thus there are many, many factorizations of A as a product of $2n-1$ 1-banded totally positive matrices.

6.3 Factorizations of totally positive matrices

What happens when L is only a lower totally positive matrix? Do we still have the above factorizations? If L is nonsingular (let us assume that it is unit diagonal) then the answer is yes, although we may lose uniqueness of the R^k and C^k. For example, if $(L)_{n,1} = (L)_{n-1,1} = 0$, then in the factorization

$$L = R^1 \cdots R^{n-1}$$

not only is $R^1 = E_{n,n-1}(\alpha_{1,n-1})$ superfluous, but in fact it may be arbitrarily chosen with sufficiently small nonnegative values for $\alpha_{1,n-1}$. (This choice will effect the entries of $R^2, \ldots, R^{n-1}$.) Nonetheless the proof of Theorem 6.6 remains valid, modulo this type of proviso, because L is nonsingular and as we have seen in Proposition 1.15 and Theorem 1.19 if

$$L\begin{pmatrix} i+1,\ldots,i+r \\ 1,\ldots,r \end{pmatrix} = 0$$

then necessarily

$$L\begin{pmatrix} j+1,\ldots,j+s \\ 1,\ldots,s \end{pmatrix} = 0$$

for all $j \geq i$ and all $s \geq r$, so that the equations in the proof of Theorem 6.6 are solvable.

However, if L is singular then we cannot apply the previous reasoning. Nevertheless, from density, and paralleling the analysis of Theorem 2.12, we have the following.

Theorem 6.10 *An $n \times n$ lower triangular matrix L is lower totally positive if and only if it can be factored in the form*

$$L = R^1 \cdots R^{n-1}$$

where R^k is a 1-banded lower totally positive matrix with $(R^k)_{i+1,i} = 0$, $i = 1, \ldots, n-k-1$. Similarly, L is lower totally positive if and only if it can be factored in the form

$$L = C^{n-1} \cdots C^1$$

where C^k is a 1-banded lower totally positive matrix with $(C^k)_{i+1,i} = 0$, $i = k+1, \ldots, n-1$.

Proof If L has either of the above factorizations, then L is a lower totally positive matrix as it is a product of the lower totally positive matrices R^k (or C^k). It remains to prove the converse claims.

A simple variation of Theorem 2.6 proves that there exist a sequence of lower strictly totally positive matrices (L_m) which approximate L, i.e., such that

$$\lim_{m \to \infty} L_m = L.$$

From Theorem 6.6 each L_m can be factored in the form

$$L_m = R_m^1 \cdots R_m^{n-1}$$

where the R_m^k are nonsingular 1-banded lower totally positive matrices with

$(R^k_m)_{i+1,i} = 0$, $i = 1, \ldots, n-k-1$. By pre- and postmultiplying by positive diagonal matrices we can factor L_m in the form

$$L_m = \widetilde{D}_m \widetilde{R}^1_m \cdots \widetilde{R}^{n-1}_m$$

where the $\widetilde{R}^k_m$ are stochastic (rows sums 1). As a consequence we also have

$$\sum_j \ell_{ij}(m) = \widetilde{d}_{ii}(m)$$

for all i, where the $\widetilde{d}_{ii}(m)$ is the (i,i) entry of $\widetilde{D}_m$. Thus all entries in each of the factors are uniformly bounded. Now that we have bounds on all the entries of the matrices in the factorization we may take a subsequence along which each of the entries of these matrices converges. Denote their respective limits by D, $R^1, \ldots, R^{n-1}$ and redefine R^1 as DR^1. Then

$$L = R^1 \cdots R^{n-1}$$

where R^k is a 1-banded lower totally positive matrix with $(R^k)_{i+1,i} = 0$, $i = 1, \ldots, n-k-1$. Similarly it follows that L can be factored in the form

$$L = C^{n-1} \cdots C^1$$

where each C^k is a 1-banded lower totally positive matrix that satisfies $(C^k)_{i+1,i} = 0$, $i = k+1, \ldots, n-1$. □

As was seen in Section 6.2, these are just two of at least 2^{n-1} (generally distinct) possible factorizations. That is, we can mix and match the factors where for each k we choose to obtain a C^k or an R^k.

Note that if L is nonsingular then each of the R^k (or C^k) must be nonsingular. Thus if the L of Theorem 6.10 is unit diagonal, then we can also make this same demand on the R^k (or C^k).

Paralleling Theorem 6.9, we can record the following.

Theorem 6.11 *Let A be an $n \times n$ totally positive matrix. Then A can be factored in many ways in the form*

$$A = L^1 \cdots L^{n-1} U^1 \cdots U^{n-1}$$

where the L^k are 1-banded, unit diagonal, lower totally positive matrices, and the U^k are 1-banded, unit diagonal, upper totally positive matrices.

We also know that we can restrict the L^k and U^k as in Theorem 6.10 so that they have a fairly large number of zero entries on their nonzero off-diagonal.

6.4 Remarks

The first factorization theorem for totally positive matrices is often called the Whitney Theorem. The name was given by Loewner [1955], although there is no evidence to indicate that Whitney [1951] ever thought of her result in this way. Whitney only talks about a "reduction theorem" and proves what we have listed as Proposition 1.9. Loewner gave this name to his factorization theorem because it is proved by repeated applications of Proposition 1.9. In fact the Whitney/Loewner method of factorization is not, from our perspective, a good factorization. Rather than eliminating successive off-diagonals, i.e., going from a $(k+1)$-banded lower (upper) totally positive matrix to a k-banded lower (upper) totally positive matrix via multiplication by one 1-banded totally positive matrix, the procedure of Whitney/Loewner eliminates (makes zero) successive columns in a lower totally positive matrix. This implies the need for $n(n-1)/2$ 1-banded factors when factoring a lower totally positive matrix (rather than just the $n-1$ factors). In the Whitney Theorem one assumes nonsingularity of the matrix. Cryer [1976] obtained 1-banded factorizations for an arbitrary totally positive matrix. But again the number of such factors can be very large. In fact Metelmann [1973] (the paper was almost unnoticed) was the first to state a factorization theorem for totally positive matrices with the correct number of factors. In Cavaretta, Dahmen, Micchelli, Smith [1981] can be found a factorization theorem (with the correct number of factors) for infinite strictly m-banded totally positive matrices. This was generalized in de Boor, Pinkus [1982] to an arbitrary totally positive nonsingular matrix (banded if the matrix is infinite). All three of these last mentioned papers, i.e., Metelmann [1973], Cavaretta, Dahmen, Micchelli, Smith [1981] and de Boor, Pinkus [1982] also consider totally positive matrices $A = (a_{ij})$ that are (r, s)-banded, i.e., for which $a_{ij} \neq 0$ only if $-s \leq i - j \leq r$, and in this case show that one can make do with $r + s$ factors (aside from the diagonal matrix). This also follows from the method of proof of the results in this chapter and Proposition 6.8. The proofs in some of the papers mentioned above tend to be rather different. The method of proof of Theorem 6.6 as presented here is basically to be found in Micchelli, Pinkus [1991]. In that paper a factorization is also given for $n \times m$ totally positive matrices. If $n > m$ then aside from $m-1$ lower and $m-1$ upper triangular $m \times m$ 1-banded totally positive factors, there are also $n-m$ totally positive factors that are $k \times (k-1)$ matrices, $k = m+1, \ldots, n$, all of whose entries are zero except for those on the two main diagonals (i.e., the (i,i)- and $(i+1,i)$-entries, $i = 1, \ldots, k-1$). This follows by expanding the

$n \times m$ totally positive matrix to an $n \times n$ totally positive matrix (just add columns of zeros), choosing the suitable factorization, and then truncating the result.

In a series of papers (see Gasca, Peña [1996] and references therein), factorizations of totally positive matrices are studied in depth using Neville elimination. These authors also use factorizations to obtain many of the characterizations of totally positive matrices as presented in Chapter 2; see e.g., Gasca, Peña [1992], [1993]. In this monograph we have chosen a different approach. Factorizations are also used in the papers of Koev [2005], Demmel, Koev [2005] and Koev [2007] where their concern is with computing using totally positive matrices. For the use of factorizations in connection with weighted planar networks, see Fomin, Zelevinsky [1999] and Fomin, Zelevinsky [2000].

Afterword

It is very difficult, if not well nigh impossible, to give an exact history of the development of any set of ideas. Nonetheless, there are four persons whose contributions "stand out" when considering the history of total positivity. They are I. J. Schoenberg, M. G. Krein, F. R. Gantmacher, and S. Karlin. Of course they did not work in a vacuum and numerous influences are very evident in their research.

It was Schoenberg who initiated the study of the variation diminishing properties of totally positive matrices in 1930 in Schoenberg [1930], and the study of Pólya frequency functions in the late 1940s and early 1950s. Independently, and unaware of Schoenberg's work, Krein was developing the theory of total positivity as it related to ordinary differential equations whose Green's functions are totally positive. Furthermore, in the mid-1930s Krein, together with Gantmacher, proved the spectral properties of totally positive kernels and matrices, and many other properties (see Gantmacher, Krein [1935], Gantmacher [1936], Gantmacher, Krein [1937], and their influential Gantmacher, Krein [1941], which was later reissued as Gantmacher, Krein [1950], and its translations in German in 1960 and in English in 1961 and 2002). These topics are the foundations upon which has been constructed the theory of total positivity. Karlin's role was somewhat different. His books Karlin, Studden [1966] and Karlin [1968], the latter titled *Total Positivity. Volume 1* (but there is no Volume 2), presented many new results and ideas and also synthesized and popularized many of these ideas.

As the reader has hopefully noted, each chapter of this monograph ends with remarks that include bibliographical references and explanations. However I wanted to take this opportunity to write a "few words" in memory of each of these gentlemen.

I. J. (Iso) Schoenberg (1903–1990) was born in Galatz, Romania and

I. J. Schoenberg, 1903–1990

died in Madison, Wisconsin. His family moved to Jassy in 1910, and he studied mathematics at the university there. He spent three years in Germany studying with both Edmund Landau and Issai Schur. Influenced by Schur, he wrote a thesis in Analytic Number Theory. (From Landau he also took Landau's daughter Charlotte (Dolli) as his first wife. He is one of many exemplars of the well-known fact that mathematical talent is an inherited trait: from father to son-in-law.) Schoenberg was interested in the problem of estimating the number of real zeros of a polynomial, and this led him to his work on variation diminishing transformations and Pólya frequency functions and kernels, which are two major topics in the theory of total positivity. A Rockefeller fellowship fortunately brought him to the United States in 1930. He spent time at the University of Chicago, was a Fellow at the newly established Princeton Institute of Advanced Studies, was on the faculty of Colby College from 1936 to 1941, and was then at the University of Pennsylvania from 1941 to 1965. In 1965 Schoenberg moved to the University of Wisconsin and joined the Mathematics Research Center and the Department of Mathematics. He retired in 1973 but remained mathematically active until his death. He contributed to many areas of mathematics, in particular total positivity and splines. For further information, see de Boor [1988] (and especially Schoenberg's autobiographical "A brief account of my life and work" at the beginning of the first volume), Askey, de Boor [1990], MacTutor, and references therein.

Mark Grigorievich Krein (1907–1989) was born in Kiev and died in Odessa. He was a truly eminent and exceedingly prolific mathematician

M. G. Krein, 1907–1989

who contributed significantly to and had a tremendous impact upon many different areas of mathematics (see e.g., Gohberg [1989] and Gohberg [1990], MacTutor, and references therein). The story of M. G. Krein, and the mathematical schools he built, is fundamentally marred by the tyranny and antisemitism to which he was constantly exposed. Krein was dismissed from his position at the University of Odessa in 1944 and from his part-time position at the Mathematical Institute of the Ukrainian Academy of Sciences in Kiev in 1952. We can only speculate on what might have been if he had been treated with the respect and dignity that were his due. In 1939 he was elected a corresponding member of the Ukrainian Academy of Sciences. He was never elected a full member. (This prompted the famous mathematical joke. *Ques:* How do you know that the Ukrainian Academy of Sciences is the best academy in the world? *Ans:* Because Krein is only a corresponding member.) From 1944 to 1954 Krein held the chair in theoretical mechanics at the Odessa Naval Engineering Institute, and from 1954 until his retirement he held the chair in theoretical mechanics at the Odessa Civil Engineering Institute. Despite being persecuted Krein received international recognition. He was elected an honorary member of the American Academy of Arts and Sciences in 1968, a foreign member of the National Academy of Sciences (of the USA) in 1979, and in 1982 he was awarded the Wolf Prize. The citation for this prize states: "Krein brought the full force of mathematical analysis to bear on problems of function theory, operator theory, probability and mathematical physics. His contributions led to important developments in the applications of mathematics to different fields ranging from theoretical mechanics to electrical engineering. His style in mathematics and his personal leadership and integrity have set standards of excellence."

F. R. Gantmacher, 1908–1964

Feliks Ruvimovich Gantmacher (1908–1964) was born in Odessa and he studied there. In 1934 he moved to Moscow, where he resided until his death. Gantmacher is, of course, known for his excellent and influential book *The Theory of Matrices* (Gantmacher [1953]), and his book with Krein; Gantmacher, Krein [1941], [1950]. He was also one of the organizers and editors of the journal *Uspekhi Mat. Nauk* (*Russian Math. Surveys*). Gantmacher was instrumental in the establishment and organization of the well-known Moscow Physico-Technical Institute, where from 1953 until his death he headed the Department of Theoretical Applied Mathematics. An obituary on Gantmacher may be found in Gantmacher [1965].

S. Karlin, 1924–2007

Samuel (Sam) Karlin (1924–2007) was born in Yonova, Poland, but he was raised in Chicago. He earned his PhD from Princeton in 1947 under the guidance of S. Bochner. From 1956 he was a faculty member at Stanford University. The breadth and depth of his interests and contributions in mathematics and in science are astounding. Karlin is the author of more than 450 papers and 10 books, and he had numerous doctoral students. Karlin was passionate about mathematics and science. His passion showed in his lectures, his lifestyle, and his interaction with students and colleagues. Karlin was widely honored, and he received honorary doctorates, numerous prizes (the John von Neumann Theory Prize in 1987, and the National Medal of Science in 1989 to name but two), and he was elected to various academies (see Karlin [2002] and MacTutor).

References

Aissen M., I. J. Schoenberg and A. M. Whitney [1952], On the generating functions of totally positive sequences I, *J. d'Anal. Math.* **2**, 93–103.

Ando T. [1987], Totally positive matrices, *Lin. Alg. and Appl.* **90**, 165–219.

Askey R. and C. de Boor [1990], In memoriam : I. J. Schoenberg (1903-1990), *J. Approx. Theory* **63**, 1–2.

Asner B. A., Jr. [1970], On the total nonnegativity of the Hurwitz matrix, *SIAM J. Appl. Math.* **18**, 407–414.

Beckenbach, E. F. and R. Bellman [1961], *Inequalities*, Springer–Verlag, New York.

Berenstein A., S. Fomin and A. Zelevinsky [1996], Parametrizations of canonical bases and totally positive matrices, *Adv. Math.* **122**, 49–149.

Boocher A. and B. Froehle [2008], On generators of bounded ratios of minors for totally positive matrices, *Lin. Alg. and Appl.* **428**, 1664–1684.

de Boor C. [1982], The inverse of totally positive bi-infinite band matrices, *Trans. Amer. Math. Soc.* **274**, 45–58.

de Boor C. [1988], *I. J. Schoenberg: Selected Papers*, 2 Volumes, Birkhäuser, Basel.

de Boor C. and A. Pinkus [1977], Backward error analysis for totally positive linear systems, *Numer. Math.* **27**, 485–490.

de Boor C. and A. Pinkus [1982], The approximation of a totally positive band matrix by a strictly banded totally positive one, *Lin. Alg. and Appl.* **42**, 81–98.

Brenti F. [1989], Unimodal, log-concave, and Pólya frequency sequences in combinatorics, *Mem. Amer. Math. Soc.* **413**.

Brenti F. [1995], Combinatorics and total positivity, *J. Comb. Theory, Series A* **71**, 175–218.

Brenti F. [1996], The applications of total positivity to combinatorics and conversely, in *Total Positivity and its Applications* eds., C. A. Micchelli and M. Gasca, Kluwer Acad. Publ., Dordrecht, 451–473.

Brown L. D., I. M. Johnstone and K. B. MacGibbon [1981], Variation diminishing transformations: a direct approach to total positivity and its statistical applications, *J. Amer. Stat. Assoc.* **76**, 824–832.

Brualdi R. A., and H. Schneider [1983], Determinantal identities: Gauss, Schur, Cauchy, Dylvester, Kronecker, Jacobi, Binet, Laplace, Muir, and Cayley, *Lin. Alg. and Appl.* **52/53**, 769–791.

Buslaev A. P. [1990], A variational description of the spectra of totally positive matrices, and extremal problems of approximation theory, *Mat. Zametki* **47**, 39–46; English transl. in *Math. Notes* **47** (1990), 26–31.

Carlson B. C. and J. L. Gustafson [1983], Total positivity of mean values and hypergeometric functions, *SIAM J. Math. Anal.* **14**, 389–395.

Carlson D. [1968], On some determinantal inequalities, *Proc. Amer. Math. Soc.* **19**, 462–466.

Carnicer J. M., T. N. T. Goodman and J. M. Peña [1995], A generalization of the variation diminishing property, *Adv. Comp. Math.* **3**, 375–394.

Cavaretta A. S. Jr., W. A. Dahmen and C. A. Micchelli [1991], Stationary subdivision, *Mem. Amer. Math. Soc.* **93**.

Cavaretta A. S. Jr., W. A. Dahmen, C. A. Micchelli and P. W. Smith [1981], A factorization theorem for banded matrices, *Lin. Alg. and Appl.* **39**, 229–245.

Crans A. S., S. M. Fallat and C. R. Johnson [2001], The Hadamard core of the totally nonnegative matrices, *Lin. Alg. and Appl.* **328**, 203–222.

Craven T. and G. Csordas [1998], A sufficient condition for strict total positivity of a matrix, *Linear and Multilinear Alg.* **45**, 19–34.

Cryer C. [1973], The LU-factorization of totally positive matrices, *Lin. Alg. and Appl.* **7**, 83–92.

Cryer C. [1976], Some properties of totally positive matrices, *Lin. Alg. and Appl.* **15**, 1–25.

Dahmen W., C. A. Micchelli, and P. W. Smith [1986], On factorization of bi-infinite totally positive block Toeplitz matrices, *Rocky Mountain J. Math.* **16**, 335–364.

Demmel J. and P. Koev [2005], The accurate and efficient solution of a totally positive generalized Vandermonde linear system, *SIAM J. Matrix. Anal. Appl.* **27**, 142–152.

Dimitrov D. K. and J. M. Peña [2005], Almost strict total positivity and a class of Hurwitz polynomials, *J. Approx. Theory* **132**, 212–223.

Edrei A. [1952], On the generating functions of totally positive sequences II, *J. d'Anal. Math.* **2**, 104–109.

Edrei A. [1953], Proof of a conjecture of Schoenberg on the generating functions of totally positive sequences, *Canadian J. Math.* **5**, 86–94.

Elias U. and A. Pinkus [2002], Non-linear eigenvalue-eigenvector problems for STP matrices, *Proc. Royal Society Edinburgh: Section A* **132**, 1307–1331.

Eveson S. P. [1996], The eigenvalue distribution of oscillatory and strictly sign-regular matrices, *Lin. Alg. and Appl.* **246**, 17–21.

Fallat S. M., M. I. Gekhtman and C. R. Johnson [2000], Spectral structures of irreducible totally nonnegative matrices, *SIAM J. Matrix Anal. Appl.* **22**, 627–645.

Fallat S. M., M. I. Gekhtman and C. R. Johnson [2003], Multiplicative principal-minor inequalities for totally nonnegative matrices, *Adv. Applied Math.* **30**, 442–470.

Fan K. [1966], Some matrix inequalities, *Abh. Math. Sem. Univ. Hamburg* **29**, 185–196.

Fan K. [1967], Subadditive functions on a distributive lattice and an extension of Szasz's inequality, *J. Math. Anal. Appl.* **18**, 262–268.

Fan K. [1968], An inequality for subadditive functions on a distributive lattice with application to determinantal inequalities, *Lin. Alg. and Appl.* **1**, 33–38.

Fekete M. and G. Pólya [1912], Über ein Problem von Laguerre, *Rend. C. M. Palermo* **34**, 89–120.

Fomin S. and A. Zelevinsky [1999], Double Bruhat cells and total positivity, *J. Amer. Math. Soc.* **12**, 335–380.

Fomin S. and A. Zelevinsky [2000], Total positivity: tests and parametrizations, *Math. Intell.* **22**, 23–33.

Friedland S. [1985], Weak interlacing properties of totally positive matrices, *Lin. Alg. and Appl.* **71**, 95–100.

Gantmacher F. [1936], Sur les noyaux de Kellogg non symétriques, *Comptes Rendus (Doklady) de l'Academie des Sciences de l'URSS* **1 (10)**, 3–5.

Gantmacher F. R. [1953], *The Theory of Matrices*, Gostekhizdat, Moscow–Leningrad; English transl. as *Matrix Theory*, Chelsea, New York, 2 vols., 1959.

Gantmacher F. R. [1965], Obituary, in *Uspekhi Mat. Nauk* **20**, 149–158; English transl. as *Russian Math. Surveys*, **20** (1965), 143–151.

Gantmacher F. R. and M. G. Krein [1935], Sur les matrices oscillatoires, *C. R. Acad. Sci. (Paris)* **201**, 577–579.

Gantmakher F. R. and M. G. Krein [1937], Sur les matrices complètement non négatives et oscillatoires, *Compositio Math.* **4**, 445–476.

Gantmacher F. R. and M. G. Krein [1941], *Oscillation Matrices and Small Oscillations of Mechanical Systems* (Russian), Gostekhizdat, Moscow-Leningrad.

Gantmacher F. R. and M. G. Krein [1950], *Ostsillyatsionye Matritsy i Yadra i Malye Kolebaniya Mekhanicheskikh Sistem*, Gosudarstvenoe Izdatel'stvo, Moskva-Leningrad, 1950; German transl. as *Oszillationsmatrizen, Oszillationskerne und kleine Schwingungen mechanischer Systeme*, Akademie Verlag, Berlin, 1960; English transl. as *Oscillation Matrices and Kernels and Small Vibrations of Mechanical Systems*, USAEC, 1961, and also a revised English edition from AMS Chelsea Publ., 2002.

Garloff J. [1982], Criteria for sign regularity of sets of matrices, *Lin. Alg. and Appl.* **44**, 153–160.

Garloff J. [2003], Intervals of almost totally positive matrices, *Lin. Alg. and Appl.* **363**, 103–108.

Garloff J. and D. Wagner [1996a], Hadamard products of stable polynomials are stable, *J. Math. Anal. Appl.* **202**, 797–809.

Garloff J. and D. Wagner [1996b], Preservation of total nonnegativity under the Hadamard products and related topics, in *Total Positivity and its Applications* eds., C. A. Micchelli and M. Gasca, Kluwer Acad. Publ., Dordrecht, 97–102.

Gasca M. and C. A. Micchelli [1996], *Total Positivity and its Applications*, eds., Kluwer Acad. Publ., Dordrecht.

Gasca M., C. A. Micchelli and J. M. Peña [1992], Almost strictly totally positive matrices, *Numerical Algorithms* **2**, 225–236.

Gasca M. and J. M. Peña [1992], Total positivity and Neville elimination, *Lin. Alg. and Appl.* **165**, 25–44.

Gasca M. and J. M. Peña [1993], Total positivity, QR factorization, and Neville elimination, *SIAM J. Matrix Anal. Appl.* **14**, 1132–1140.

Gasca M. and J. M. Peña [1995], On the characterization of almost strictly totally positive matrices, *Adv. Comp. Math.* **3**, 239–250.

Gasca M. and J. M. Peña [1996], On factorizations of totally positive matrices, in *Total Positivity and its Applications* eds., C. A. Micchelli and M. Gasca, Kluwer Acad. Publ., Dordrecht, 109–130.

Gasca M. and J. M. Peña [2006], Characterizations and decompositions of almost strictly totally positive matrices, *SIAM J. Matrix Anal.* **28**, 1–8.

Gladwell G. M. L. [1998], Total positivity and the QR algorithm, *Lin. Alg. and Appl.* **271**, 257–272.

Gladwell G. M. L. [2004], Inner totally positive matrices, *Lin. Alg. and Appl.* **393**, 179–195.

Gohberg I. [1989], Mathematical Tales, in *The Gohberg Anniversary Collection*, Eds. H. Dym, S. Goldberg, M. A. Kaashoek, P. Lancaster, pp. 17–56, Operator Theory: Advances and Applications, Vol. 40, Birkhäuser Verlag, Basel.

Gohberg I. [1990], Mark Grigorievich Krein 1907–1989, *Notices Amer. Math. Soc.* **37**, 284–285.

Goodman T. N. T. [1995], Total positivity and the shape of curves, in *Total Positivity and its Applications* eds., C. A. Micchelli and M. Gasca, Kluwer Acad. Publ., Dordrecht, 157–186.

Goodman T. N. T. and Q. Sun [2004], Total positivity and refinable functions with general dilation, *Appl. Comput. Harmon. Anal.* **16**, 69–89.

Gross K. I. and D. St. P. Richards [1995], Total positivity, finite reflection groups, and a formula of Harish-Chandra, *J. Approx. Theory* **82**, 60–87.

Holtz O. [2003], Hermite-Biehler, Routh-Hurwitz and total positivity, *Lin. Alg. and Appl.* **372**, 105–110.

Horn R. A. and C. R. Johnson [1991], *Topics in Matrix Analysis*, Cambridge University Press, Cambridge.

Karlin S. [1964], The existence of eigenvalues for integral operators, *Trans. Amer. Math. Soc.* **113**, 1–17.

Karlin S. [1965], Oscillation properties of eigenvectors of strictly totally positive matrices, *J. D'Analyse Math.* **14**, 247–266.

Karlin S. [1968], *Total Positivity. Volume 1*, Stanford University Press, Stanford, CA.

Karlin S. [1972], Some extremal problems for eigenvalues of certain matrix and integral operators, *Adv. in Math.* **9**, 93–136.

Karlin S. [2002], Interdisciplinary meandering in science. 50th anniversary issue of Operations Research. *Oper. Res.* **50**, 114–121.

Karlin S. and A. Pinkus [1974], Oscillation properties of generalized characteristic polynomials for totally positive and positive definite matrices, *Lin. Alg. and Appl.* **8**, 281–312.

Karlin S. and W. J. Studden [1966], *Tchebycheff Systems: with Applications in Analysis and Statistics*, Interscience Publishers, John Wiley, New York.

Katkova O. M. and A. M. Vishnyakova [2006], On sufficient conditions for the total positivity and for the multiple positivity of matrices, *Lin. Alg. and Appl.* **416**, 1083–1097.

Kellogg O. D. [1918], Orthogonal function sets arising from integral equations, *Amer. J. Math.* **40**, 145–154.

Kellogg O. D. [1929], *Foundations of Potential Theory*, Springer-Verlag, Berlin.

Kemperman J. H. B. [1982], A Hurwitz matrix is totally positive, *SIAM J. Math. Anal.* **13**, 331–341.

Koev P. [2005], Accurate eigenvalues and SVDs of totally nonnegative matrices, *SIAM J. Matrix. Anal. Appl.* **27**, 1–23.

Koev P. [2007], Accurate computations with totally nonnegative matrices, *SIAM J. Matrix. Anal. Appl.* **29**, 731–751.

Koteljanskii D. M. [1950], The theory of nonnegative and oscillating matrices (Russian), *Ukrain. Mat. Z.* **2**, 94–101; English transl. in *Amer. Math. Soc. Transl., Series 2* **27** (1963), 1–8.

Koteljanskii D. M. [1955], Some sufficient conditions for reality and simplicity of the spectrum of a matrix, *Mat. Sb. N.S.* **36**, 163–168; English transl. in *Amer. Math. Soc. Transl., Series 2* **27** (1963), 35–42.

Krein M. G. and A. A. Nudel'man [1977], *The Markov Moment Problem and Extremal Problems*, Transl. Math. Monographs, Vol. 50, Amer. Math. Society, Providence.

Kurtz D. C. [1992], A sufficient condition for all the roots of a polynomial to be real, *Amer. Math. Monthly* **99**, 259–263.

Loewner K. [1955], On totally positive matrices, *Math. Z.* **63**, 338–340.

Lusztig G. [1994], Total positivity in reductive groups, in *Lie Theory and Geometry*, Progress in Mathematics, **123**, Birkhäuser, Boston, 531–568.

Maló E. [1895], Note sur les équations algébriques dont toutes les racines sont réelles, *Journal de Mathématiques Spéciales* **4**, 7–10.

Marden M. [1966], *Geometry of Polynomials*, Math. Surveys, Vol. 3 (2nd edition), Amer. Math. Society, Providence.

Markham T. L. [1970], A semi-group of totally nonnegative matrices, *Lin. Alg. and Appl.* **3**, 157–164.

Marshall A. W. and I. Olkin [1979], *Inequalities: Theory of Majorization and its Applications*, Academic Press, New York.

Metelmann K. [1973], Ein Kriterium für den Nachweis der Totalnichtnegativität von Bandmatrizen, *Lin. Alg. and Appl.* **7**, 163–171.

Micchelli C. A. and A. Pinkus [1991], Descartes systems from corner cutting, *Constr. Approx.* **7**, 161–194.

Motzkin Th. [1936], *Beiträge zur Theorie der linearen Ungleichungen*, Doctoral Dissertation, Basel, 1933. Azriel Press, Jerusalem.

Mühlbach G. and M. Gasca [1985], A generalization of Sylvester's identity on determinants and some applications, *Lin. Alg. and Appl.* **66**, 221–234.

Pinkus A. [1985a], Some extremal problems for strictly totally positive matrices, *Lin. Alg. and Appl.* **64**, 141–156.

Pinkus A. [1985b], n-widths of Sobolev spaces in L^p, *Const. Approx.* **1**, 15–62.

Pinkus A. [1985c], *n-Widths in Approximation Theory*, Springer–Verlag, Berlin.

Pinkus A. [1996], Spectral properties of totally positive kernels and matrices, in *Total Positivity and its Applications* eds., C. A. Micchelli and M. Gasca, Kluwer Acad. Publ., Dordrecht, 477–511.

Pinkus A. [1998], An interlacing property of eigenvalues of strictly totally positive matrices, *Lin. Alg. and Appl.* **279**, 201–206.

Pinkus A. [2008], Zero minors of totally positive matrices, *Electronic J. Linear Algebra* **17**, 532–542.

Pitman J. [1997], Probabilistic bounds on the coefficients of polynomials with only real zeros, *J. Comb. Theory, Ser. A* **77**, 279–303.

Pólya G. and G. Szegő [1976], *Problems and Theorems in Analysis II*, Springer–Verlag, New York.

Rahman Q. I. and G. Schmeisser [2002], *Analytic Theory of Polynomials*, Oxford University Press, Oxford.

Schoenberg I. J. [1930], Über variationsvermindernde lineare Transformationen, *Math. Z.* **32**, 321–328.

Schoenberg I. J. and A. Whitney [1951], A theorem on polygons in n dimensions with application to variation-diminishing and cyclic variation-diminishing linear transformations, *Compositio Math.* **9**, 141–160.

Schumaker L. L. [1981], *Spline Functions: Basic Theory*, John Wiley & Sons, New York.

Shapiro B. Z. and M. Z. Shapiro [1995], On the boundary of totally positive upper triangular matrices, *Lin. Alg. and Appl.* **231**, 105–109.

Shohat J. A. and J. D. Tamarkin [1943], *The Problem of Moments*, Math. Surveys, Vol. 1, Amer. Math. Society, Providence.

Skandera M. [2004], Inequalities in products of minors of totally nonnegative matrices, *J. Alg. Comb.* **20**, 195–211.

Smith P. W. [1983], Truncation and factorization of bi-infinite matrices, in *Approximation Theory, IV* (College Station, Tex., 1983), 257–289, Academic Press, New York.

Stieltjes T. J. [1894–95], Recherches sur les fractions continues, *Annales Fac. Sciences Toulouse* **8**, 1–122; **9**, 1–47.

Wagner D. [1992], Total positivity of Hadamard products, *J. Math. Anal. Appl.* **163**, 459–483.

Wang Y. and Y.-N. Yeh [2005], Polynomials with real zeros and Pólya frequency sequences, *J. Comb. Theory, Ser. A* **109**, 63–74.

Whitney A. [1952], A reduction theorem for totally positive matrices, *J. d'Anal. Math.* **2**, 88–92.

Author index

Subject index